The Contextualized UML

Practical Use of Unified Modeling Language in Japanese Industry

by

Dr. Atsushi Tanaka

authorHOUSE

1663 LIBERTY DRIVE, SUITE 200
BLOOMINGTON, INDIANA 47403
(800) 839-8640
www.authorhouse.com

First published by AuthorHouse 06/08/04

ISBN: 1-4184-2324-6 (e)
ISBN: 1-4184-2325-4 (sc)

Library of Congress Control Number: 2004105429

Printed in the United States of America
Bloomington, Indiana

This book is printed on acid-free paper.

Dedication

To Almighty God, my family and people who promote world peace

Acknowledgement

First of all, my adoration, indebtedness and glorification are humbly submitted to the Almighty God for seeing me through.

I wish to highly acknowledge the absolutely necessary help, encouragement and moral support of brother Rev./Dr. Samuel B. Kamara. The title of my book was decided after getting inspiration from his book "The Contextualized Gospel" which was published by AuthorHouse. I recommend Sam's book to people who want to know the approaches of the contextualization in details.

I am very grateful to the following, who have helped me generously in one way or the other: Mr. and Mrs. Nagahara, who had set up a time and place for our deliberation, brother Robin White, who designed the front and back covers of this book, everyone at 1stBooks, especially sister Becky Dulin, who is the author services representative. I am also thankful to Dr. Jim Rumbaugh, who is one of the designers of UML. He answered two of my difficult questions.

It goes without saying that I am grateful to my parents for everything.

List of Figures

Table of Contents

PART ONE
BRIEF INTRODUCTION TO THE UNIFIED MODELING LANGUAGE

PART TWO
BUSINESS MODELING CONTEXTUALIZATION

Vendor Managed Inventory

PART THREE
INDUSTRIAL SYSTEM CONTEXTUALIZATION

Preface

Nowadays, Japan has been established by electronics and system technologies and had the second/third big economic power in the world. In almost all areas of modern life, computer-based systems are indispensable. However, such systems suffer from their insufficient correctness or reliability so that software development projects can fail. Through arguments for development problems, the field of software engineering has appeared by means of adapting traditional engineering methods to software development. Software engineering is related not only to developers but also users. Thus, social changes and problems have a great influence on software engineering.

Since foreign affiliated firms are advancing their technologies in the Japanese field, there is the problem of communication between this foreign personnel and the Japanese. Japanese language is difficult for foreign people, just as the English is difficult for Japanese. Even though the Esperanto artificial language was developed for communication purpose, the cost for mastering it is too high. There is need for the introduction of another artificial language in graphical representation which can be mastered by foreigners and Japanese people easily. The author therefore, in this book, discusses the Unified Modeling Language, for short UML.

This language has developed as a language suitable for object orientation in software engineering. Instead of a natural language, UML gives strict description. As a formal language, UML is used for analysis and design. This is why UML is said to be a semi-formal language.

Although graphic-based modeling techniques are of increasing interest, the specifications of UML are quite large. However, all of them are always not necessary for practical applications. The practical essentials of UML are presented in the book with straightforward English and simple examples. It is not assumed that readers have special knowledge on computers. Thus, the book is appropriate to begin to learn about UML. Many diagrams are included in the book. The fact (in the form of illustrations) shows the powerful communication capacity of UML.

This book consists of three parts. Part one gives a brief introduction to UML. It covers the basics of UML and shows how they can be used. It has nine chapters and can also be used as a reference to the most common parts of UML. In chapter 1, use case diagrams for user and system requirements are discussed. Chapter 2 describes class diagrams which indicate sets of objects and their relationships. In chapter 3, we discuss interaction diagrams which are used as behavioral snap shots. The actions of each object are described by state machine diagrams shown in chapter 4. Chapter 5 gives activity diagrams for system or business process. As shown in chapter 6, implementation diagrams are used for arranging other diagrams. In chapter 7, we discuss Petri net diagrams which are supposed to be included in UML. Contrary to popular belief, Petri nets have been used to develop industrial systems. Petri nets have sub classes such as state machines and event graphs. Chapter 8 describes timed event graphs which are used for performance evaluation. In chapter 9, we discuss how to design system architecture based on Petri nets. After reading part one, you can go into more details with part two and three. These parts are practical and experience-based guides.

Part two describes business modeling contextualization with UML. Due to contextualization, you can find the practical use of UML for business modeling. Chapter 10 describes retail business modeling. In chapter 11, wholesale business modeling are discussed. Through these chapters, business processes as well as modeling techniques by UML are presented. These business processes and modeling techniques are described, based on a practical point of view. Then, chapter 12 describes scientific approaches to business modeling and shows effective methods for practitioners.

Part three describes industrial system contextualization with UML. The contextualization intends to bridge the gap between analysis and design process. UML turns out to be applied not only to business systems, but also engineering systems through contextualization. In chapter 13, requirement and system analysis processes are discussed by a development example of a factory automation equipment. Chapter 14 describes design process of the system given by the previous chapter. Although these chapters gives general ideas for developing industrial systems, new social problems have occurred. We should tackle the problems with a paradigm shift from object orientation to agent orientation. Chapter 15 discusses Petri net-based approaches toward a next generation. Petri nets have come a long way, especially, the skyrocketing object-oriented Petri nets. Based on this idea, an agent-oriented modeling is discussed for next generation industrial systems.

PART ONE

BRIEF INTRODUCTION TO THE
UNIFIED MODELING LANGUAGE

Chapter 1. Use Case Diagram

Use Case and Actor

We develop a system to improve a business environment. A development plan from a user point of view is necessary in order to develop a system that the user requires. A **use case model**[1] is a model that defines system requirements from the user view point with model elements such as an **actor**, a **use case** and their **relationships**.

An actor expresses a user, an external system or a business process, which is indicated by a doll icon called **stickman**. A use case is a set of functionalities that a user participates in, which is indicated by an **ellipse**. A development domain, or a system is associated to a **rectangle** in which use cases are located. The relationship between two model elements such as an actor and a use case is indicated by a **solid line** connecting them.

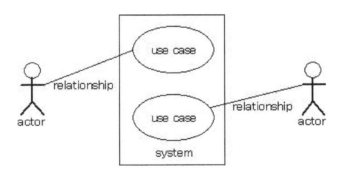

Figure 1.1 Use Case Diagram

Business Use Case

First, in general, we should find actors, then use cases are to be considered. The granularity of use cases is important. An appropriate use case is a use case whose

3

purpose is achieved only by **completing the execution without interruption**. Such a use case, as shown in Fig.1.2, is particularly said to be a **business use case[2]**. An actor collaborating with a business use case is called a **business actor[2]**.

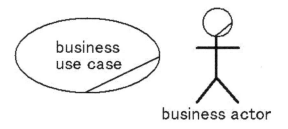

Figure 1.2 Business Use Case and Actor

For example, in the following rental business use cases, to lend and to accept a return are appropriate.

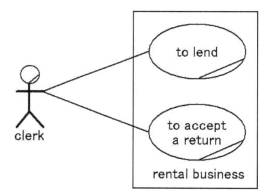

Figure 1.3 Rental Business Use Case

System Use Case

A business use case could be divided into smaller use cases called **system use cases**. In the rental business, there could be system use cases such as in checking a member card or calculating a rental fee; It is not appropriate as a business description but used to express the functional details in a business use case.

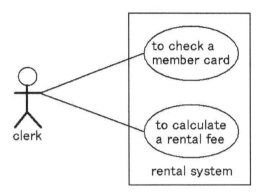

Figure 1.4 Rental System Use Case

Use Case Description

The details of a business use case are expressed by a **use case description**. A use case description consists of an **abstract**, a **scenario** and an **event flow** in an informal way.

An abstract is the compact explanation of a use case. For example, in a use case to join a member, the abstract could be written as *"In a rental shop, a customer has to join the member first to rent an item"*.

A scenario, which is a concrete business description, is divided into the main scenario where the purpose of the use case is achieved, and into an exceptional scenario where it is not achieved.

For example, in a use case to join a member on a Web site, the main scenario is described as follows.

A customer, Sam, pushes a button to join a member of a Web rental shop; The member registration is displayed. Sam inputs his name "Samuel B. Kamara" with the telephone number 12-3456-7890, e-mail address Sam@example.com and the credit card number 1234-5678-9012-3456; When he pushes a button to register, the Web system checks the input data and registers Sam's data to a member list. Finally, his member ID and password will be displayed.

Figure 1.5 Use Case Scenario

An event flow description based on some scenarios, is a general description of scenarios with pre-condition. It must be satisfied before the flow execution; The post-condition must be satisfied after the flow execution and a candidate flow should be used instead of the main flow.

For example, in a use case to join a member on a Web site, the event flow is described as follows.

Pre-condition:
A customer is not a member.
Main flow:
The customer pushes a button to join a member, then member registration is displayed. So, the customer inputs the name, telephone number, e-mail address, credit card number and so on.
Candidate flow:
If input data are wrong, the customer inputs them again.
Post-condition:
The customer is a member.
Exceptional flow:
If the credit card number is on a blacklist, "The credit card could not be used" is displayed.

Figure 1.6 Use Case Event Flow

Relationship

The types of use case relationship are **association**, **inclusion**, **extension** and **generalization**.

Association

An association is used between an actor and a use case, implying their collaboration, which does not ensure the successful input or output, indicated by a **solid line**. An association could have the **multiplicity**, the number of actors or use cases.

For example, in the following use case, the actor side notation – 1..3 means that the number of clerks is from 1 to 3. The use case side notation * means that the number of checks is from 0 to infinity.

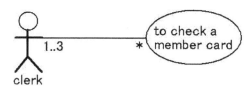

Figure 1.7 Association with Multiplicity

Inclusion

An inclusion is used between two use cases. If use cases A and B share a part of their event flows, the common part is defined as a use case C. Then, the inclusion connecting A and B to C is indicated by a **dashed arrow with a stick arrowhead** and a notation stereo type **<<include>>**. This arrow "A and B include C". It implies that "A and B depend on C". Thus, a **dependency** like this is indicated by a **dashed arrow with a stick arrowhead**.

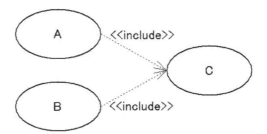

Figure 1.8 Use Case Inclusion

For example, in the following Web shopping use cases, a customer has to login the Web site in advance to use the search or buy function.

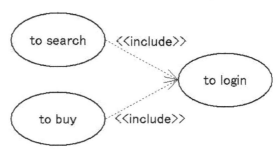

Figure 1.9 Web Shopping Functional Inclusion

Extension

An extension is used between two use cases. A use case event flow has an optional part. The part is separated as an **extension use case**. An extended use case, which is called a **basis use case**, has an **extension point** explaining the selected functional part. An extension use case is connected to the basis use case with a **dashed arrow with a stick arrowhead** and a notation stereo type **<<extend>>**.

8

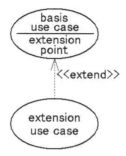

Figure 1.10 Use Case Extension

For example, in the following Web shopping use cases, a customer can select how to pay. A credit card is one possible option.

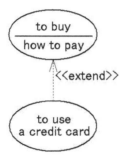

Figure 1.11 Web Shopping Functional Extension

Generalization

A generalization is used between either two actors or two use cases. The former is the generalization of actors and the latter is the generalization of use cases. If a set of use cases with which an actor A, collaborating, is a sub set of use cases with which another actor B, collaborating, then B is said to inherit use cases from A. The generalization of B is indicated by a **solid arrow with an unfilled arrowhead**.

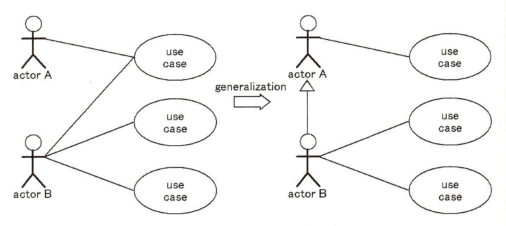

Figure 1.12 Actor Generalization

For example, in the following rental system actors, a manager gets a stock in addition to jobs which a clerk does.

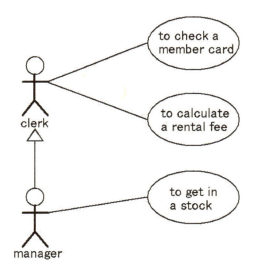

Figure 1.13 Rental System Actor Generalization

Actors are grouped by an **abstract actor**, whose name is written in inclined characters, by using a generalization. For example, in the following actors, a regular member and a fellow member are grouped by a member.

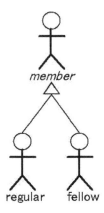

Figure 1.14 Abstract Actor in terms with Membership

If a use case B has an event flow of a use case A with an additional B event flow description, then B is said to inherit an event flow from A. The generalization of B is indicated by a **solid arrow with an unfilled arrowhead**.

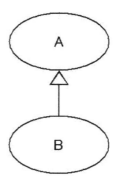

Figure 1.15 Use Case Generalization

For example, in the following member registration use cases, a customer can be a special member as well as a regular member.

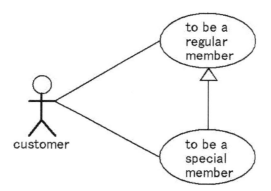

Figure 1.16 Member Registration Use Case Generalization

Use cases are grouped by an **abstract use case**, whose name is written in inclined characters, by using a generalization. For example, in the following use cases, to buy a Christmas gift and to buy a birthday gift are grouped under "to buy a gift".

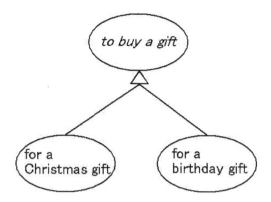

Figure 1.17 Gift Buy Use Case Generalization

Chapter 2. Class Diagram

Object and Class

UML is used to model an **object-oriented system**, that is, a system that has to do with collaborating objects. An object is characterized by its **name, states** and **behaviors**[1]. For example, if a system to do with 5000 students is developed, the system consists of the corresponding 5000 objects. However, it is not efficient to model a number of objects.

Thus we model a set of objects that are said to be a **class** instead of modeling objects[2]. A class is characterized by its **name, attributes** (as state variables) and **operations** (as behaviors). According to the class definition, an objects collaboration exhibits a behavior with an object A which instructs another object B to execute the operation, called the **message passing**.

A class is indicated by a rectangle consisting of three compartments, where the operation part is omitted and only the class name is expressed, as shown in Fig.2.1.

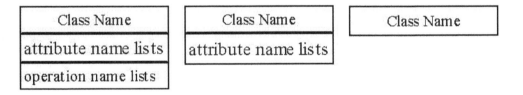

Figure 2.1 Class Notation

In the middle compartment, attributes are written by the following format:
visibility attribute name: type name = initial value

Where the visibility, **public** (accessible from anywhere), **protected** (accessible from subclasses), **private** (accessible from the class itself inside) and **package** (accessible from a package discussed later), are expressed by +, #, - and

15

~ for short respectively, are fully defined. For the type name, **int** (integer), **string**, etc. are also defined. The initial value is a value given to an object when the object is created from the class.

In the bottom compartment, operations are written by the following format:

visibility operation name(argument name: type): return type

The argument is in a value that the operation receives a message by passing it in and starting the execution. The return type is a type of value that the operation returns to the instruction object, in completing the execution.

Relationship

The types of class relationship are **association** including **aggregation** and **composition (composite aggregation)**; **Dependency** includes **realization** and **generalization**.

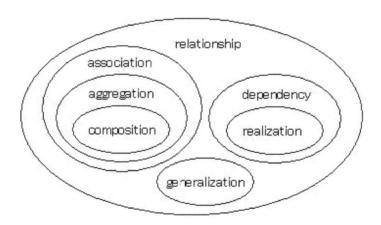

Figure 2.2 Class Relationship

Association

An association, indicated by a **solid line**, expresses an enabled event between classes. The occurrence of the event means that a message passes from a class object to another class object. An association could have the **association name** with a **filled triangle** indicating the direction to read. The **role** or **multiplicity** of each class determines the meaning or number of the class objects, respectively.

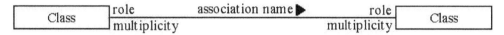

Figure 2.3 Class Association

For example, as shown in Fig.2.4, an employee belongs to a department. The employee is a member of the department, and the department is at a working location for the employee. The number of the employees is from 1 to infinity. The number of the department is from 0 to 1, which is at most 1.

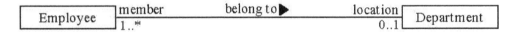

Figure 2.4 Association between Employee and Department

As shown in Fig.2.5, the association implies that the passing direction of a message between two classes is undefined.

Figure 2.5 Association with Undefined Message Passing

If an one-way message passing is defined, the association is indicated by **a solid arrow with a stick arrowhead**. For example, as shown in Fig.2.6, it is possible for an employee to access a supervisor.

Figure 2.6 Association with One-way Message Passing

Similarly, as shown in Fig.2.7, a two-way message passing is also explicitly indicated.

Figure 2.7 Association with Two-way Message Passing

Aggregation

An aggregation is a kind of association. When a class object consists of another class object, such an association between the whole class and the portion class is an aggregation. Then an **unfilled diamond-shaped symbol** is attached to the whole side terminal – the aggregation. For example, as shown in Fig.2.8, the member list consists of members.

Figure 2.8 Class Aggregation

Composition

A composition (composite aggregation) is an aggregation where the life cycle of the whole class is the same as that of the portion class. Then a **filled diamond-shaped symbol** is attached to the whole side terminal – the composition. For example, as shown in Fig.2.9, the manufacturing line is in full operation of machines.

Figure 2.9 Class Composition

Dependency

A dependency discussed in the previous chapter is a **use association**, indicated by a **dashed arrow with a stick arrowhead**. This use association has three types of association:

(i) As an **argument** of a class operation, another class object is used.

(ii) In a class operation said to be **local**, anther class object is used.

(iii) From any class object said to be **global**, another class object is used.

Figure 2.10 Class Dependency

Generalization

If a class B has the attributes and operations of a class A and additional attributes and operations, then B is said to inherit the attributes and operations from A. A is the generalization of B, which is indicated by a **solid arrow with an unfilled arrowhead**. As shown in Fig.2.11, A is called a **superclass** and B is called a **subclass**.

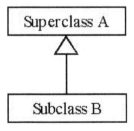

Figure 2.11 Class Generalization

For example, as shown in Fig.2.12, the member class is the generalization of the regular and fellow member classes.

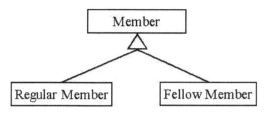

Figure 2.12 Member Class Generalization

The common operations of subclasses are defined in an **abstract class** that is a superclass, whose name is written in **inclined characters**. For example, as shown in Fig.2.13, the common operation of each window subclass "close" is defined in the abstract superclass. In the superclass, a concrete process of the inclined characters operation *"open"* called an **abstract operation**, is not defined, because the behavior of "open" is different in each subclass. Thus, the process of "open" is defined in each subclass. Naturally, the subclasses inherit "close" from the superclass.

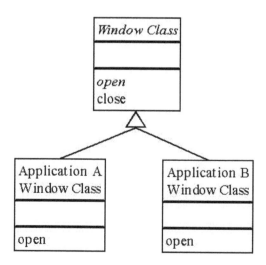

Figure 2.13 Abstract Class in terms with Window System

Realization

A realization is a kind of dependency through a generalization. A dependency means that a class depends on another class. However, the operations of the other class which has the dependence is not obvious.

So, the depended operations are defined as abstract operations in an abstract superclass. Such an abstract class having only abstract operations is called an **interface**. A generalization by an interface, actually meaning a dependency, is said to be a **realization**; It is indicated by a **dashed arrow with an unfilled arrowhead** and a notation stereo type **<<interface>>** in the interface.

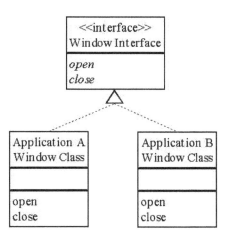

Figure 2.14 Interface in terms with Window System

An interface is expressed by a circle for short. This symbol is called a **lollipop**. For example, as shown in Fig.2.15, two application classes A and B have a dependency to the lollipop.

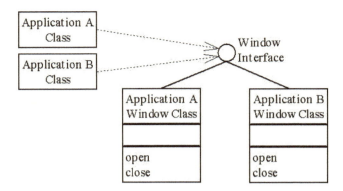

Figure 2.15 lollipop Interface

Reflexive Association

When a class has an association with itself, it is expressed by a **reflexive association**. For example, as shown in Fig.2.16, the employee class is expressed by one class and a reflexive association.

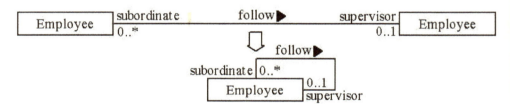

Figure 2.16 Reflexive Association

Association Class

To take and use data about an association between two classes, the data are stored into an **association class**. For example, as shown in Fig.2.17, while an employee uses a high performance computer, the use record data are stored into the association class Use Record. By the operations "take" and "use", the data are stored and used.

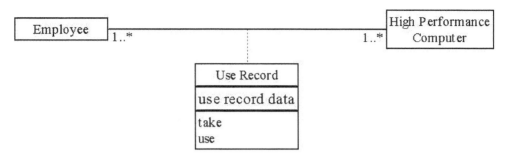

Figure 2.17 Association Class

N-ary-association

Associations among more than two classes are said to be a **N-ary-association**, where N is the number of classes, indicated by an **unfilled diamond-shaped symbol** at a cross point on the associations. For example, as shown in Fig.2.18, a 3-ary-association is expressed among a employee, a department and a country.

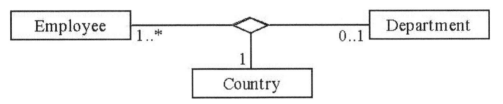

Figure 2.18 3-ary-association

Parameterized Class

A **parameterized class** is a special class in order to generate usual classes. For example, as shown in Fig.2.19, the List class is a parameterized class and its parameter is T. T is defined as Customer by a notation stereo type **<<bind>> (Customer)**, which is said to be an **explicit binding**. By this, the Customer List class is generated.

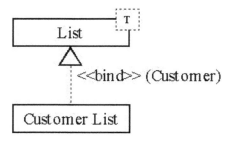

Figure 2.19 Parameterized Class by Explicit Binding

Also by **implicit binding**, the Customer List class is expressed as shown in Fig.2.20.

Figure 2.20 Parameterized Class by Implicit Binding

Constraint

A **constraint** is written between { and }. For example, as shown in Fig.2.21, the balance of the Bank Account class is more than 1000$.

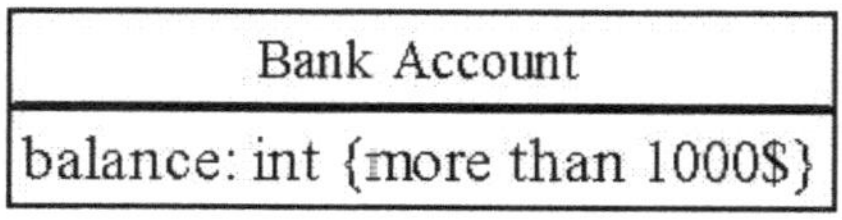

Bank Account
balance: int {more than 1000$}

Figure 2.21 Attribute Constraint

A constraint **{ordered}** means that the data of attributes are ordered. For example, as shown in Fig.2.22, the data of the Member class are ordered by ascending member ID.

Figure 2.22 Constraint {ordered}

A constraint **{Xor}** means that the only one of aggregation is effective. For example, as shown in Fig.2.23, the Member List class has either regular members or fellow members.

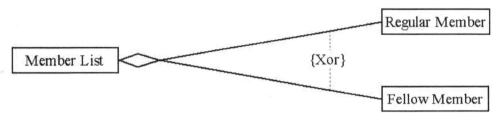

Figure 2.23 Constraint {Xor}

Constraints on generalization are defined such as **{overlapping}**, **{disjoint}**, **{complete}**, **{incomplete}** and so on. The {overlapping} means that a superclass object is able to have more than one subclass object. And the {disjoint} means that a superclass object cannot have more than one subclass object. The {complete} means that a subclass cannot be added. And the {incomplete} means that a subclass can be added.

For example, as shown in Fig.2.24, one subclass object, that is, one item is effective by the {disjoin} and a new item can be added by the {incomplete}.

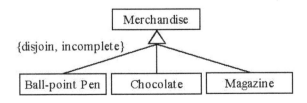

Figure 2.24 Constraint {disjoin, incomplete}

Qualifier

A **qualifier** is a constraint on the multiplicity of aggregation. For example, as shown in Fig.2.25, using the qualifier of a member ID for the Member List class, the multiplicity for the Member class changes from 0..* to 0..1.

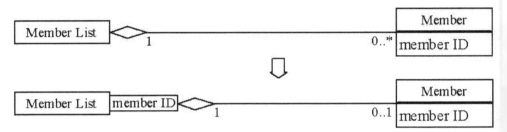

Figure 2.25 Qualifier

Chapter 3. Interaction Diagram

Object Diagram

An object has a lifecycle from the creation to the destruction. Collaborating objects at an instantaneous time are described by an **object diagram**[1]. An object is indicated by the following rectangles in which the name of either the object, the object and the class, or the only class called a **no-name object**, is underlined. Also, a concrete actor is similarly expressed by its underlined name.

Figure 3.1 Object and Concrete Actor Notation

A relationship among objects is said to be a **link**. For example, as shown in Fig.3.2, Mike has a chocolate and a magazine in his shopping cart on a Web shopping system.

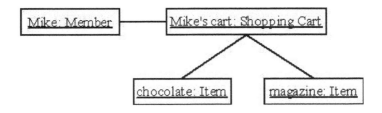

Figure 3.2 Shopping Cart Object Diagram

An object diagram added a ball-point pen to the shopping cart is shown as in Fig.3.3.

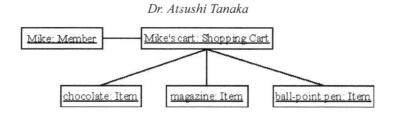

Figure 3.3 Changed Shopping Cart Object Diagram

An object notation could have the data of attributes, as shown in Fig.3.4, if necessary.

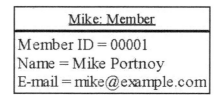

Figure 3.4 Object with Attribute Data

Communication Diagram

A **communication diagram**[2] expresses a communication among objects or actors described by an object diagram with message passing. A passing of message is indicated by an arrow. This means directing of the passing on which a **message label** is an instruction to execute an operation; A **sequence number** which is a message order, is written. The message label format is as follows:

sequence number: return value: =message name(argument)

As shown in Fig.3.5, there are three types of arrows. A solid arrow with a filled solid arrowhead means a **synchronous message**. A solid arrow with stick arrowhead means an **asynchronous message**. A dashed arrow with stick arrowhead means a **return message**.

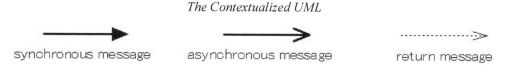

synchronous message   asynchronous message   return message

Figure 3.5 Arrow Type

For example, as shown in Fig.3.6, a customer actor sends an asynchronous message "display" to the item window object at first. Secondly, a synchronous message is sent from the item window object to the item control object. Thirdly, a synchronous message is sent from the item control object to the item data object. Then, the return message data "item" is returned from the item data object to the item control object, and the data "item" is returned from the item control object to the item window object. Finally, a reflexive message "display" of the item window object is executed.

Note that the item window object is created from a presentation class; The item control object is created from a supervisor class and the item data object is created from a plant class, which will be discussed later.

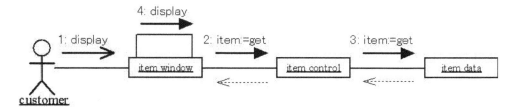

Figure 3.6 Communication

Each message of a concurrent sequence is identified by appended alphabet characters. For example, as shown in Fig.3.7, the messages of the #3a, #3b and #3c are concurrent.

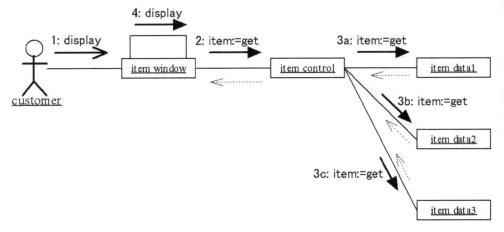

Figure 3.7 Concurrent Communication

A **predecessor** / is used when a message cannot be sent till another message passing is completed. For example, as shown in Fig. 3.8, the message of the #**3a, 3b/3** cannot be sent till the messages passing of the #3a and #3b are completed.

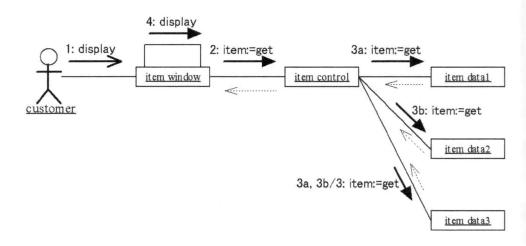

Fig.3.8 Communication with Predecessor

An object working autonomously is called an **active object**[3]. It is indicated by a **bold rectangle**. And a non-active object is called a **passive object**. Also, a class creating an active object is said to be an **active class**. For example, as shown in Fig.3.9, the search robot on the Web autonomously gets the Web Pages 1, 2 and 3.

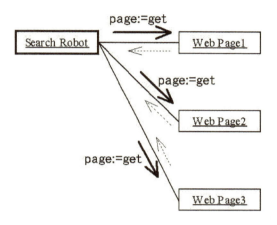

Figure 3.9 Active Object

Sequence Diagram

A **sequence diagram**[4] expresses a communication among objects or actors according to the time sequence without links. For example, the sequence of Fig 3.6 is described by the sequence diagram of Fig.3.10. A dashed vertical line under each actor or object is said to be a **lifeline** by the existence of the corresponding actor or object. In addition, a lifeline is also determined by a time evolution from up to down. Thus, this example presents the sending of messages according to the order of sequence. The sequence number may be omitted.

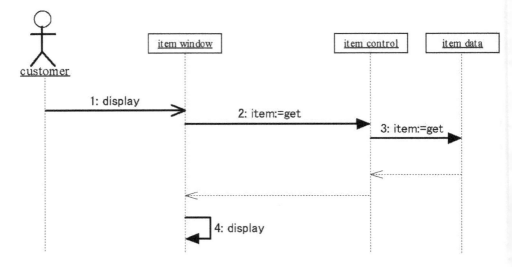

Figure 3.10 Sequence Diagram

Activation

The execution time of an operation is said to be an **activation**, indicated by an unfilled rectangle on the corresponding lifeline. For example, as shown in Fig.3.11, there are activations by "1: display" and "4: display" for the item window object, "2: item:=get" for the item control object and "3: item:=get" for the item data object.

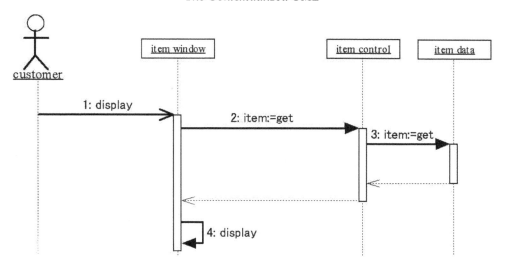

Figure 3.11 Activation

Branching Lifeline

A sequence diagram may be described by a no-name object, that is, a class; A lifeline could have a branch which makes a possibility for another lifeline. For example, as shown in Fig.3.12, the Register Control class may send two kinds of messages "[regular] register" to the Regular Member class and "[fellow] register" to the Fellow Member class, where such an expression between [and] is said to be a **guard condition**. This type of message is able to be sent only if a guard condition is satisfied. An object of the Member List class could receive two kinds of messages from an object of the Regular Member class or the Fellow Member class; Thus the lifeline of the Member List class has a branch.

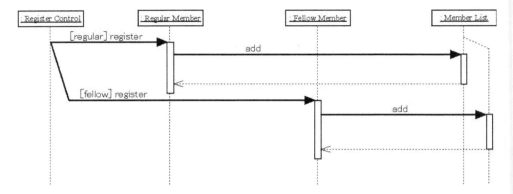

Figure 3.12 Branching Lifeline

Creating and Destroying Object

The creation of an object is expressed by a message passing, and the destruction of an object is indicated by a **big X**. For example, as shown in Fig.3.13, the item window object is created by the message passing from the customer actor. Its destruction is indicated by the big X. Note that there is no lifeline under the big X.

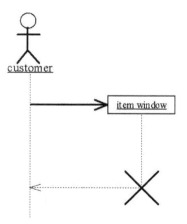

Figure 3.13 Creating and Destroying Object

Chapter 4. State Machine Diagram

State Machine

A **state machine diagram**[1] expresses possible states of an object and their transitions. As shown in Fig.4.1, a state of an object is indicated by a **round-corner rectangle**.

Figure 4.1 State Notation

As shown in Fig.4.2, the creation of an object is expressed by an **initial pseudo state** indicated by a **filled circle**. The event of the creation transitioning to an **initial state** is indicated by a **solid arrow with stick arrowhead**.

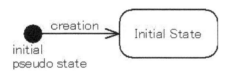

Figure 4.2 Creating Object

As shown in Fig.4.3, the destruction of an object is expressed by a **final state** indicated by a **filled double circle**.

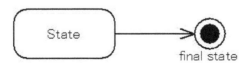

Figure 4.3 Destroying Object

35

If a state has some internal states, such a state is said to be a **composite state** or a **super state**. A composite state, which is a hierarchical expression of a state machine, has a **sub machine** whose state is said to be a **sub state**. For example, as shown in Fig.4.4, the order of states in the execution is A, B, C and D.

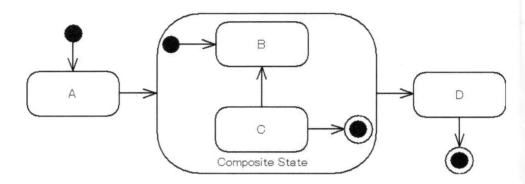

Figure 4.4 Composit State

Mealy Machine

The event description format is as follows:

event name [guard condition] /action

Where an event is enabled only if the guard condition is satisfied, the event occurs, then an action is executed in the state transition. A state machine described in this way is said to be a **Mealy machine**.

For example, as shown in Fig.4.5, at the initial state "Stop", if a signal is "blue", then the action "go" is executed. At the state "Go", the action "stop" is executed if a signal is "red", and the action "return" is executed if a signal is "yellow". At the state "Return", if a signal is "red", then the action "stop" is executed.

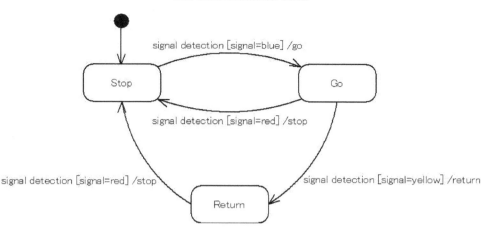

Figure 4.5 Mealy Machine

Moore Machine

As shown in Fig.4.6, a state in details consists of a **state name compartment** which shows a state name, and an **internal transition compartment** which shows an **entry action**, an **activity** and an **exit action**.

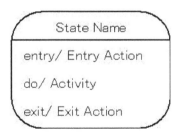

Figure 4.6 State Description

Getting into a state, its entry action is executed. And getting out a state, its exit action is executed. An activity is executed from entry to exit or till completing the execution. A state machine described in this way is said to be a **Moore machine**. Generally, a Moore machine is larger than a Mealy machine. The choice depends on the context. As shown in Fig.4.7, the Moore machine of Fig. 4.5 is obtained.

37

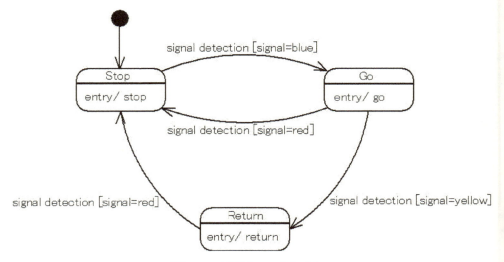

Figure 4.7 Moore Machine

Completion Transition

Regardless of the occurrence of an event, a state transition due to the completion of an action or an activity is said to be a **completion transition** or a **trigger-less transition**. This transition has no event name, but a guard condition could be attached. For example, as shown in Fig.4.8, a completion transition from the state "Timer" to the state "Stop" is executed by the completion of the "count down".

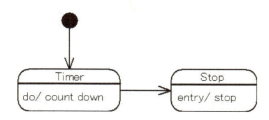

Figure 4.8 Completion Transition

History Pseudo State

A state of a sub machine is recorded in a **history pseudo state**, indicated by a **circled H**. So, even exiting from a sub machine whose sub state is the current state, the sub machine can be executed at the sub state again by entering the history pseudo state of the sub machine.

For example, as shown in Fig.4.9, the state indicated by the arrow directed to the history pseudo state is the initial sub state – "Cooling". By the occurrence of the event "off", the execution transitions from the composite state "Air-conditioning" to the state "Stop", and the state of the sub machine is recorded in the history pseudo state. By the occurrence of the event "on", the execution which transitions to the sub state recorded in the history pseudo state of the Air-conditioning, is made possible.

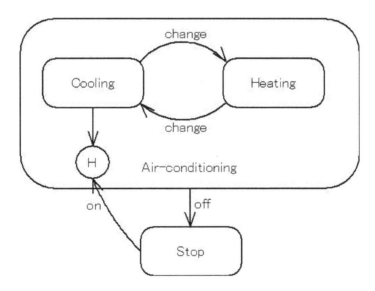

Figure 4.9 History Pseudo State

When a sub state of a composite state has a sub machine, whose sub state is also recorded in history pseudo state, such a history is said to be a **deep history**

pseudo state. It is indicated by a **circled H***, while the history of Fig. 4.8 is said to be a **shallow history pseudo state**.

For example, as shown in Fig.4.10, the sub states are "Cooling" and "Heating" in a composite state "Air-conditioning". Each sub state has a sub machine whose sub states are "Strong" and "Weak". In order to record a current state such as the "Strong" or "Weak", the deep history pseudo state is used.

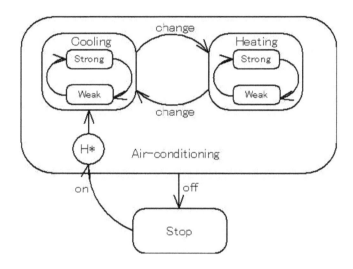

Figure 4.10 Deep History Pseudo State

Concurrent Sub State

A state is partitioned into more than one sub state transition by a **dashed line**. Such sub states are said to be **concurrent sub states**, while non-concurrent sub states are said to be **sequential sub states**. For example, as shown in Fig.4.11, the composit state "Process" has three state transitions which are executed concurrently.

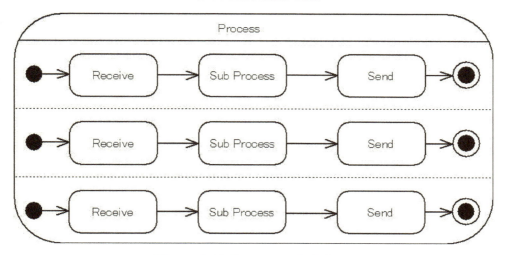

Figure 4.11 Concurrent Sub State

State Transition Analysis

It is not always easy to model state machines. So in general when we model a state machine, the corresponding **state transition table** is used to **analysis** and **test** a relationship between a state and an event.

For example, we consider an autonomous robot[2] which moves to a destination found by itself automatically within a certain time, whose constraint is a **soft real-time**. As shown in Fig.4.12, we model three states "Stop", "Slow" and "Fast" as working modes where the "Stop" is the initial state.

Figure 4.12 Modeling State

41

If there is no obstacle in the mode "Stop", as shown in Fig.4.13, the robot starts moving in the mode "Slow".

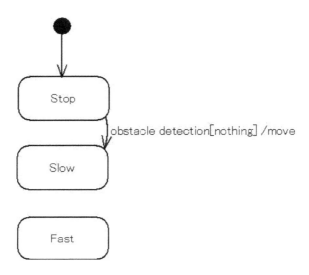

Figure 4.13 Starting to Move

If the robot reaches a destination, as shown in Fig.4.14, the robot stops.

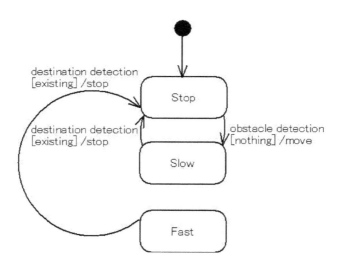

Figure 4.14 Stopping

If the remaining time is short in the mode "Slow", as shown in Fig.4.15, the robot starts moving in the mode "Fast".

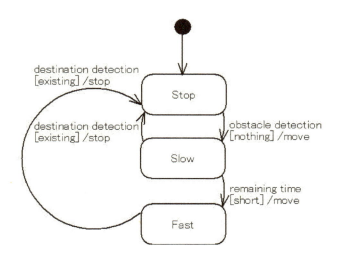

Figure 4.15 Moving Fast

If the time is over, as shown in Fig.4.16, a message "time over" is displayed.

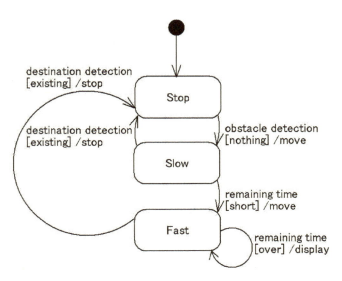

Figure 4.16 Time Over

Dr. Atsushi Tanaka

As shown in Fig.4.17, a Moore machine of Fig. 4.16 is obtained.

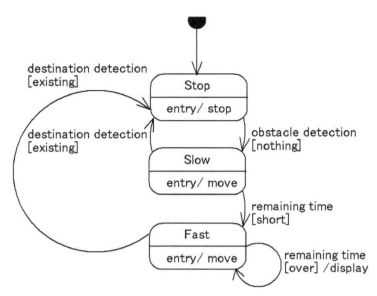

Figure 4.17 Moore Machine

As shown in Fig.4.18, a state transition table of Fig. 4.17 is derived.

Event \ State	Stop entry/ stop	Slow entry/ move	Fast entry/ move
obstacle detection [nothing]	go to Slow		
remaining time [short]		go to Fast	
remaining time [over]			display "time over" go to Fast
destination detection [existing]		go to Stop	go to Stop

Figure 4.18 State Transition Table

From the table, when the remaining time is over in the mode "Slow", a message "time over" should be displayed and the robot starts moving in the

44

mode "Fast". The transition "go to Fast" is redundant for the state "Fast", thus the transition is deleted. Moreover, **X** indicating **disabled** is written in the empty cell for the event "destination detection [existing]", and / indicating **neglected** is written in each empty cell for the other events. Therefore, as shown in Fig.4.19, the table is modified.

Event \ State	Stop entry/ stop	Slow entry/ move	Fast entry/ move
obstacle detection [nothing]	go to Slow	/	/
remaining time [short]	/	go to Fast	/
remaining time [over]	/	display "time over" go to Fast	display "time over"
destination detection [existing]	X	go to Stop	go to Stop

Fig.4.19 Modified State Transition Table

According to the table of Fig. 4.19, as shown in Fig.4.20, the state machine is modified.

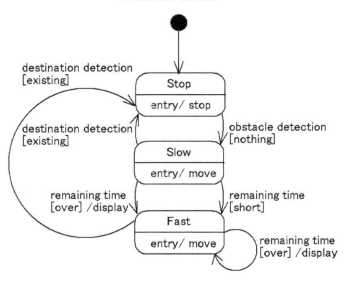

Figure 4.20 Modified State Machine

State Transition Extension

In a practical point of view, we consider a state transition table at first, rather than a state machine. The technique of state machines is **state-based analysis**, while that of state transition table is **event-based analysis** like Petri nets which will be discussed later. Thus, as shown in Fig.4.21, only two states "Stop" and "Move" are enough for modeling events in a state transition table.

Event \ State	Stop entry/ stop	Move entry/ move
obstacle detection [nothing]	change to Slow go to Move	/
remaining time [short]	/	change to Fast
remaining time [over]	/	display "time over" change to Fast
destination detection [existing]	X	go to Stop

Figure 4.21 Event-based State Transition Table

As shown in Fig.4.22, a state machine of Fig.4.21 is obtained.

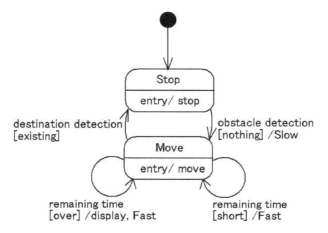

Figure 4.22 Simplified State Machine

Now, when an obstacle approaches the stopping robot, the robot changes the mode to the state "Check". In the mode "Check", as shown in Fig.4.23, if there is no obstacle in front, then the robot start moving fast, otherwise the robot retains the mode "Stop".

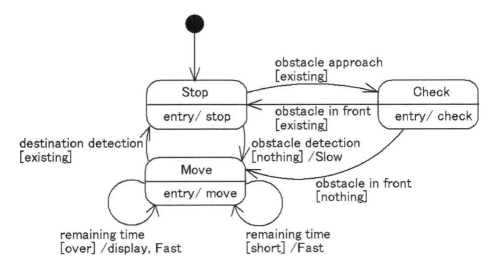

Fig.4.23 State-based Extension of State Machine

As shown in Fig.4.24, a state transition table of Fig. 4.23 is derived.

Event \ State	Stop entry/ stop	Move entry/ move	Check entry/ check
obstacle detection [nothing]	change to Slow go to Move	/	
remaining time [short]	/	change to Fast	
remaining time [over]	/	display "time over" change to Fast	
destination detection [existing]	X	go to Stop	X
obstacle approach [existing]	go to Check	/	/
obstacle in front [nothing]	/	/	change to Fast go to Move
obstacle in front [existing]	/	/	return to Stop

Figure 4.24 Extended State Transition Table

For the mode "Check" in the table, the mode is returned to the "Stop" for the "obstacle detection [nothing]". The mode is changed to the "Fast" for the "remaining time [short]" or "remaining time [over]". A message "time over" is displayed for the "remaining time [over]". Therefore, as shown in Fig.4.25, the table is modified.

State Event	Stop entry/ stop	Move entry/ move	Check entry/ check
obstacle detection [nothing]	change to Slow go to Move	/	return to Stop
remaining time [short]	/	change to Fast	change to Fast
remaining time [over]	/	display "time over" change to Fast	display "time over" change to Fast
destination detection [existing]	X	go to Stop	X
obstacle approach [existing]	go to Check	/	/
obstacle in front [nothing]	/	/	change to Fast go to Move
obstacle in front [existing]	/	/	return to Stop

Figure 4.25 Modified State Transition Table

Finally, as shown in Fig.4.26, a completed state machine is obtained from the table of Fig.4.25.

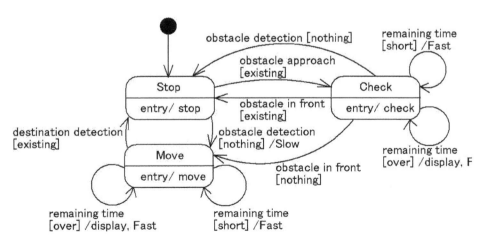

Figure 4.26 Completed State Machine

Chapter 5. Activity Diagram

Activity Diagram

An **activity diagram**[1] expresses some procedures by an **activity state**. An activity state such as an **initial pseudo state**, a **final state**, a **guard condition**, etc. is like a state machine. Compared to a state machine, an activity state expresses a process, as shown in Fig.5.1, indicated by a more slow curve.

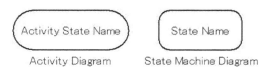

Figure 5.1 Activity State Notation

Fork and Join

A procedural **fork** or **join** is indicated by a **filled rectangle**. For example, as shown in Fig.5.2, the paying process and the selling process as well as stock taking are parallel after the fork. After that, the shipping and receiving processes are serial by the join transaction.

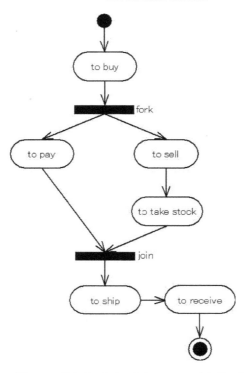

Figure 5.2 Activity Diagram with Fork and Join

Decision and Merge

A fork or joining with a guard condition is said to be a **decision** or a **merge** respectively. It is indicated by a **diamond-shaped symbol**. For example, as shown in Fig.5.3, an item value is discounted if a customer has a discount ticket.

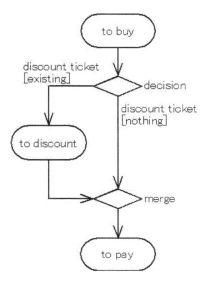

Figure 5.3 Decision and Merge

Signal

An activity communicating with another activity is expressed by an **object flow (data flow)**; An **object flow** is indicated by a **dashed arrow with a stick arrowhead**; Between **signal sending** and **receiving**, there is a description by a **convex** and **concave pentagon** respectively. For example, as shown in Fig.5.4, a signal URL is sent from a Web browser to a Web server.

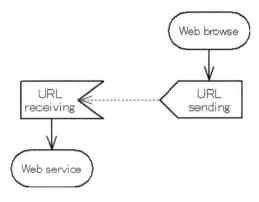

Figure 5.4 Signal Sending and Receiving

Partition

A state of an activity is expressed in a lane called a **partition**. For example, as shown in Fig.5.5, there are three partitions such as a customer to buy, pay and receive, a salesclerk to sell and take stock, and a shipping clerk to ship.

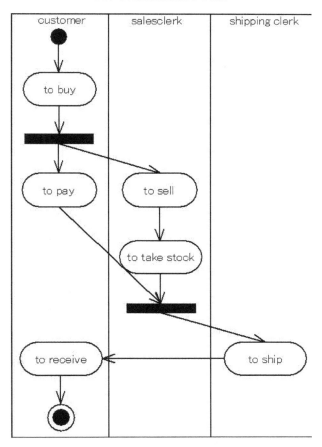

Figure 5.5 Partitioning Activity

Object

An object, with an object flow (data flow) indicated by a **dashed arrow with a stick arrowhead**, is described in an activity diagram. For example, as shown in Fig.5.6, the object "order sheet" is described on the flow from the fork to the activity to sell.

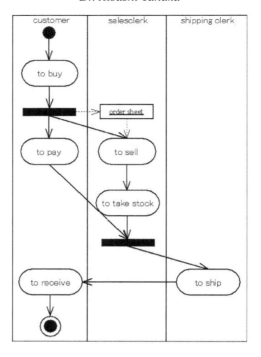

Figure 5.6 Object in Activity Diagram

Interaction Overview Diagram

An activity diagram using a sequence diagram, in an **interaction frame**, instead of an active state, is said to be an **interaction overview diagram**[2]. An interaction frame, which is used as a sub system in a sequence diagram as well as an interaction overview diagram, is labeled by an **operator** such as **sd**, **ref**, etc. The sd indicates a sequence diagram and the ref indicates a reference to an interaction on another diagram. For example, as shown in Fig.5.7, two test cases are done parallel. Note that a **timing diagram**[3] could be used for describing timing constraints on an interaction, however, **timed Petri nets**[4], which will be discussed later, are more useful.

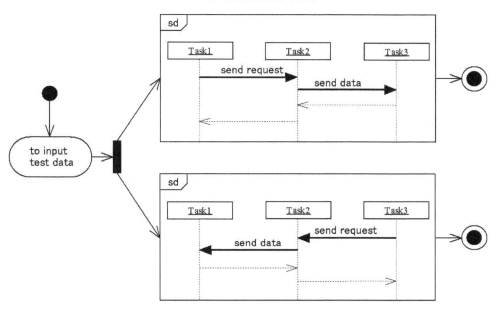

Figure 5.7 Interaction Overview Diagram for System Test

Chapter 6. Implementation Diagram

Component Diagram

A **component diagram**[1] expresses components and their dependency. As shown in Fig.6.1, a component, which is a **sub system** or a **physical software** such as a source file, an executable file, a dynamic link library, a database table, a document, etc., is described. These software are indicated by notations stereo type **<<source>>**, **<<executable>>**, **<<DDL>>**, **<<table>>** and **<<document>>**, respectively. From a practical point of view, a **deployment diagram**[2] is illustrated and will be discussed later. A component, whose name is **underlined**, is used in a deployment diagram.

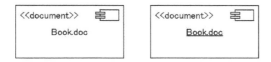

Figure 6.1 Component Notation

Dependency of components implies that a component is used to **compile**, **interpret**, or **execute** other components. For example, as shown in Fig.6.2, the Supervisor.class is executed by interpreting the Registration.html. At the same time, the memberList.class and the member.class are similarly executed.

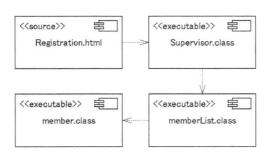

Figure 6.2 Component Dependency

59

Naturally, a dependency is also used for an interface called a lollipop. For example, as shown in Fig.6.3, the update of the memberDatabase is only used by the member.class.

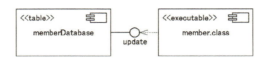

Figure 6.3 Interface Dependency

Deployment Diagram

A **deployment diagram**[2], as shown in Fig.6.4, expresses an architecture of hardware called **nodes** in which an instance of a component may be put. From a practical point of view, a node, whose name is **underlined** with its **type name**, is used.

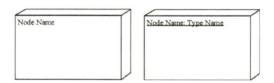

Figure 6.4 Node Notation

As shown in Fig.6.5, when the testRegistration.html is interpreted by the Web client Internet Explorer, the Supervisor.class, etc. are executed on the Web application server Tomcat.

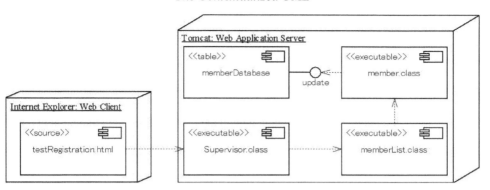

Figure 6.5 Deployment Diagram

Composit Structure Diagram

To reuse software parts for the sake of an efficient development, the parts structure is described by a **composit structure diagram**[3]. For example, as shown in Fig.6.6, we consider the composit aggregation.

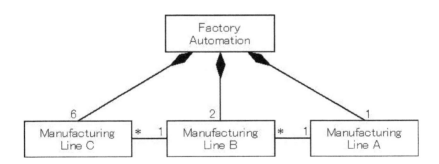

Figure 6.6 Composit Aggregation

Using a composit structure diagram, a class is hierarchically decomposed into an internal structure in order to take a complex object and break it down into parts. As shown in Fig.6.7, the composit aggregation example, implying that the factory automation consists of three kinds of manufacturing lines A, B and C, is also described by the composit structure diagram. The internal classes which

61

the class Factory Automation has are called **parts**. It is easier to understand the composite structure diagram than the composite aggregation for the sake of software reuse.

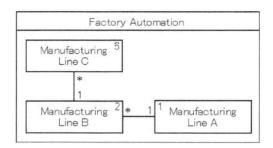

Figure 6.7 Composit Structure of Class

A service of a class is expressed by a **port** indicated by an **unfilled square**. As shown in Fig.6.8, the internal class Manufacturing Line A has the port labeled by parts which implies that parts are transferred from the Manufacturing Line A to the Manufacturing Line B.

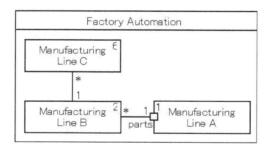

Figure 6.8 Internal Port of Composit Structure

Furthermore, as shown in Fig.6.9, the class manufacturing Line A is decomposed into the internal classes Machine A, B and C. By this composit structure, it is implied that three machines A, B and C work together and parts are carried out from the C.

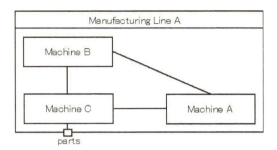

Figure 6.9 External Port of Composit Structure

Now, as shown in Fig.6.10, we consider another example. The multimedia player component has two kinds of interfaces which are expressed by a **lollipop** and a **socket** indicated by a **circle** and a **half circle** respectively.

The lollipops controller and converter are used by other classes. Thus in a class diagram, such interfaces are written with a notation stereo type **<<Provided Interface>>**.

The sockets channel and media used are corresponding with other lollipops. Thus in a class diagram, such interfaces are written with a notation stereo type **<<Required Interface>>**.

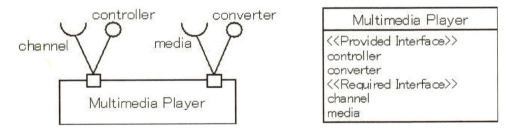

Figure 6.10 Component and Class

As shown in Fig.6.11, the component Multimedia Player is described by the composit structure diagram. A connection between an internal component called a **part** and a **port** are expressed by a **delegation connector**. A delegation connector is directed from a lollipop to a part, or from a part to a socket, indicated by a **solid arrow with a stick arrowhead**.

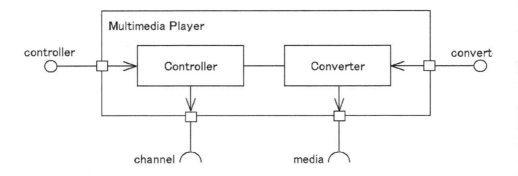

Figure 6.11 Composit Structure of Component

Package Diagram

A composite structure diagram is a **runtime grouping**, while a **compile-time grouping** is, as shown in Fig.6.12, expressed by a **package diagram**[4], which can include every model of UML.

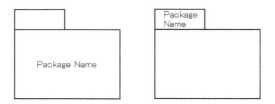

Figure 6.12 Package Notation

As shown in Fig.6.13, the System in the package Bank has the Repository included in the package Plant, and the Controller and the Transaction Facade grouped in the package Supervisor.

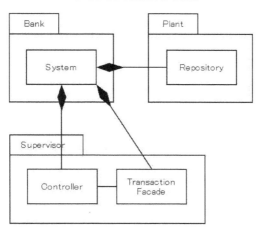

Figure 6.13 Bank System

As shown in Fig.6.14, the relationship among the Presentation, the Supervisor and the Plant is **dependency**. On the other hand, the relationship that the Presentation, the Supervisor and the Plant are in the Bank is **hierarchy**.

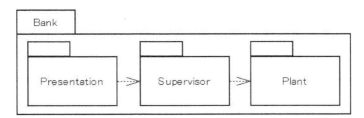

Figure 6.14 Package Dependency and Hierarchy

A hierarchy of packages can be expressed by an **inclusion icon** indicated by a **circled cross**. By this expression, as shown in Fig.6.15, the Bank system example is also described hierarchically.

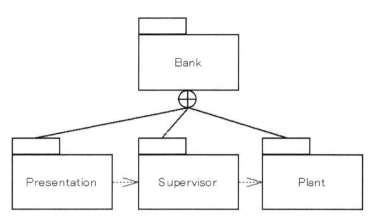

Figure 6.15 Hierarchical Package

Chapter 7. Petri Net Diagram

Discrete Event System

The **object orientation** has been proposed for **discrete event simulation**[1] in the field of **discrete event systems (DESs)**[2]. A DES is a system, consisting of a set of discrete state and a set of discrete events, that a state transitions to another state by the occurrence of an event. An enabled event is decided by a state, and a state transition is driven by the occurrence of an event. Thus a DES evolves by repetitions of such state transitions. Examples of DESs are **man-made systems** such as computer systems, communication systems, manufacturing systems, traffic control systems, database management systems and so on. It is not easy to deal with DESs whose complex natures are **asynchronism, concurrency, non-determinism**, etc.

Asynchronism is a nature in which an event occurs independently of other events[3].

Concurrency is a nature where any event may occur earlier than other events in a set of events which are not ordered[3].

Non-determinism is a nature that a state has which can transition to more than one state by an event[3].

Modeling DES

Many models have been proposed for DESs. Especially, models for describing dynamics are classified into **action models, logical models, algebraic models, probabilistic models**, etc[4]. Also, these models are classified into **timed models** and **untimed models**. Untimed models are used for **qualitative aspects** such as **reachability, liveness**, etc. Timed models are used for **quantitative aspects** such as **cycle, throughput**, etc.

Petri nets are **weighted directed bipartite graphs** consisting of two kinds of nodes indicating a state and an event[5]. A graphically described Petri net is able to execute its simulation. Also, due to its graph structure and algebraic expression, the asynchronous, concurrent and non-deterministic model can be analyzed and synthesized.

A Petri net itself is a simple dynamical model based on state transitions. Thus, according to the subject of modeling, many modified or extended models have been proposed. System theory on Petri nets is said to be **net theory**. Net theory has been proposed by C. A. Petri since 1962 and developed as a field of computer science. Nowadays net theory forms a branch of system theory.

Hybrid Dynamical System

Petri nets can express systems consisting of only **discrete variables**, however, real-world systems have **continuous variables** as well as discrete variables[6,7,8]. For example, as shown in Fig.7.1, water is transferred from the well to the tank by the pump system. In case where the well is empty or the tank is filled with water, the pump stops. Then, a pump action switching on/off is a discrete event described by discrete variables, while the increasing or decreasing amount of water is a continuous event described by continuous variables. Such a system where discrete and continuous variables are mixed is said to be a **hybrid dynamical system**[9].

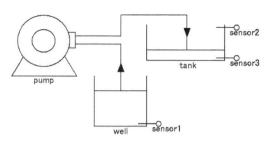

Figure 7.1 Hybrid Dynamical System

Hybrid dynamical systems widely exist in the world and are recently watched by system and computer scientists and engineers with keen interest. Many models have been proposed for hybrid dynamical systems, above all, **hybrid Petri nets**[6,7,8] discussed later are quite significant.

Place/Transition Net

The fundamental model of Petri nets is a **place/transition net**[10] consisting of two kinds of nodes such as a **place** indicating a local state and a **transition** indicating an event, indicated by a **circle** and a **filled rectangle** respectively. A place/transition net is an event-based model. As shown in Fig.7.2, the modeling approach is assembling a **net module**[11] where a transition has **weighted directed arcs** from **input places** or to **output places**.

Dr. Atsushi Tanaka

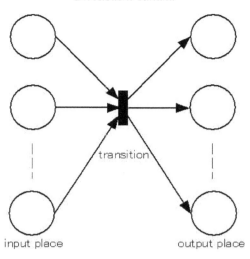

Figure 7.2 Net Module

Data in a place are said to be **tokens** indicated by **filled circles** or a **figure** indicating the number of tokens, that is, a token implies whether a condition is satisfied or not, or the amount of data. If Z^+ is a set of positive integers, then

[Definition 7.1] Place/transition nets are defined by

$$PN =< P,T,I,O,M_0 >$$

where P is a finite set of places, T is a finite set of transitions, $P \cap T = \phi$ and $P \cup T \neq \phi$. $I : P \times T \rightarrow Z^+ \cup \{0\}$ denotes a weight of an input arc of a transition. $O : P \times T \rightarrow Z^+ \cup \{0\}$ denotes a weight of an output arc of a transition. M_0 is an initial marking which is a distribution of tokens over places.

The occurrence of an event associated with a transition is said to be **firing**. By firing of a transition, a token is taken away from an input place, and a token is added to an output place. A transition without input places is said to be a **source transition**. A source transition unconditionally fires because it does not have input places indicating the enabling condition of an event. A transition without output places is said to be a **sink transition**. Obviously, a sink transition only consumes tokens in input places.

Next, firing rules of a place/transition net are presented where $^\bullet T_j$ or $T_j^\bullet$ is a set of input or output places of a transition T_j. By a marking function $M : P \rightarrow Z^+ \cup \{0\}$, the number of tokens in a place P_i is indicated by $M(P_i)$. Let the number of places be $n = |P|$, then a marking is expressed by a n-ary row vector where $M(P_i)$ is the i-th element of the vector. Thus a marking is equivalent to a marking function, then both are written by M .

[Definition 7.2] Firing rules of place/transition nets are defined as follows.

Step1. A transition T_j is called **firing-enabled** if it is a source transition or $M(P_i) \geq I(P_i,T_j)$ for each input place $P_i \in {}^\bullet T_j$.

Step2. When a firing-enabled transition T_j fires, $I(P_i,T_j)$ tokens are taken away from each input place $P_i \in {}^\bullet T_j$ and $O(P_i,T_j)$ tokens are added to each output place $P_i \in T_j^\bullet$.

For example, as shown in Fig.7.3, in the marking $M = (2,2,0)$ of the upper diagram, the number of tokens in the input place P_1 is equal to the weight of the input arc $I(P_1,T_1) = 2$. And the number of tokens in the input place P_2 is greater than the $I(P_2,T_1) = 1$. Thus T_1 is firing-enabled. When T_1 fires, $I(P_1,T_1)$ tokens are taken away from P_1, $I(P_2,T_1)$ token is taken away from P_2 and $O(P_3,T_1)$ tokens are added to P_3, resulting in the marking $M' = (0,1,2)$ of the lower diagram.

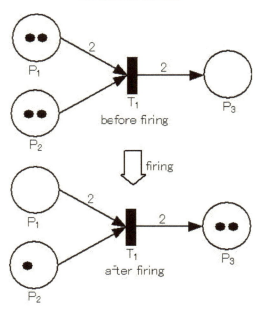

Figure 7.3 Firing of Transition

At last, as shown in Fig.7.4, the pump dynamics of Fig.7.1 are described by a place/transition net.

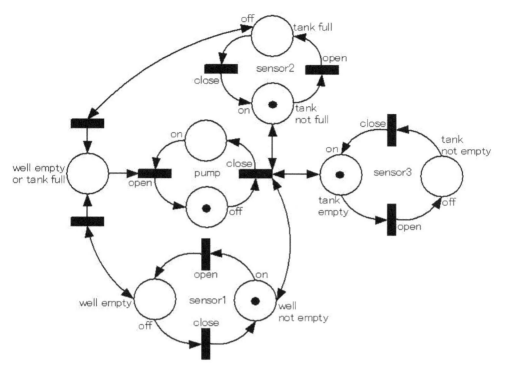

Figure 7.4 Place/Transition Net of Pump Dynamics

Hybrid Petri Net

Hybrid Petri nets are extended Petri nets consisting of **discrete Petri nets** modeling DESs and **continuous Petri nets** modeling continuous systems[6,7,8]. As shown in Fig.7.5, hybrid Petri nets have nodes such as a discrete place (D-place) and a discrete transition (D-transition) for discrete variables, and a continuous place (C-place) and a continuous transition (C-transition) for continuous variables.

Figure 7.5 Node of Hybrid Petri Net

If Z^+ is a set of positive integers and R^+ is a set of positive real numbers, then

[Definition 7.3] Hybrid Petri nets are defined by
$$HPN =< P,T,I,O,h,M_0 >$$

where P is a finite set of places, T is a finite set of transitions, $P \cap T = \phi$ and $P \cup T \neq \phi$. $h : P \cup T \rightarrow \{D,C\}$ is a hybrid function that indicates whether each node is associated with a discrete variable or a continuous variable. $I : P \times T \rightarrow R^+ \cup \{0\}$ denotes a weight of an input arc of a transition. If $h(P_i) = D$, a constraint $I(P_i,T_j) \in Z^+ \cup \{0\}$ is added. $O : P \times T \rightarrow R^+ \cup \{0\}$ denotes a weight of an output arc of a transition. If $h(P_i) = D$, a constraint $O(P_i,T_j) \in Z^+ \cup \{0\}$ is added. When $h(P_i) = D \wedge h(T_j) = C$, an arc connecting P_i and T_j is said to be a **permission arc** such that $I(P_i,T_j) = O(P_i,T_j)$. A permission arc allows firing of a transition T_j if a token exists in the D-place P_i, and forbids firing of T_j otherwise, where tokens in P_i are never changed due to firing of T_j. M_0 is an initial marking which is a non-negative integer for a D-place, or a non-negative real number for a C-place.

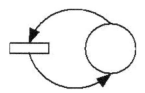

Figure 7.6 Permission Arc between C-transition and D-place

As shown in Fig.7.7, pump and well dynamics of Fig.7.1 are described by a hybrid Petri net where the amounts of water in the well and tank are l and k respectively.

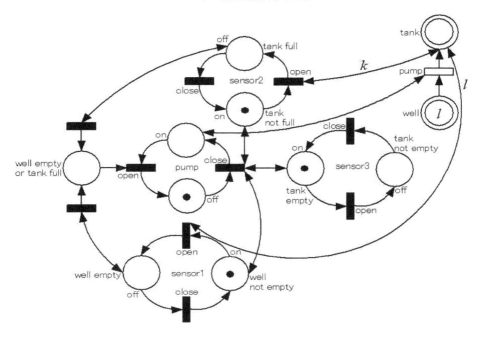

Figure 7.7 Hybrid Petri Net of Pump and Well Dynamics

Continuous dynamics depend on the concept of time. Then hybrid Petri nets have to be introduced the information of time. Thus timed hybrid Petri nets have been proposed[6,7,8].

[Definition 7.4] Timed Hybrid Petri nets are defined by
$$THPN =< HPN, Tempo, Delay >$$

where $Tempo : T \times R^+ \rightarrow R^+$ is a function assigning a positive real number to each C-transition T_j, and $Tempo(T_j, t) = V_j(t)$ indicates the maximum instantaneous firing speed of T_j at time t. $Delay : P \rightarrow R^+ \cup \{0\}$ is a function assigning a non-negative real number to each place P_i, and $Delay(P_i) = d_i$ is a delay time of a token supplied to P_i by firing of the input transition until it can contribute to firing of the output transition.

We introduce delay between firing decision and actual firing. The use of tokens for a certain time after firing is forbidden. Hence, marking is expressed as $M = M^a + M^u$ where M^a and M^u denote an available marking and an unavailable marking at the time of firing.

[Definition 7.5] Firing rules of a D-transition are defined as follows:

Step1. A D-transition T_j is called **firing-enabled** at time t if it is a source transition or $M^a(P_i) \geq I(P_i, T_j)$ for each input place $P_i \in {}^\bullet T_j$.

Step2. If T_j firing-enabled at time t fires, $I(P_i, T_j)$ available tokens are taken away from each input place $P_i \in {}^\bullet T_j$ and $O(P_i, T_j)$ unavailable tokens are added to each output place $P_i \in T_j^\bullet$.

Step3. The unavailable tokens added to each output place $P_i \in T_j^\bullet$ remain unavailable during the delay time of d_i. These unavailable tokens become available after d_i is passed.

[Definition 7.6] Firing rules of a C-transition are defined as follows:

Step1. A C-transition T_j is called **strongly enabled** at time t if it is a source transition or $M^a(P_i) \geq I(P_i, T_j)$ for each input D-place $P_i \in {}^\bullet T_j$ and $M^a(P_i) > 0$ for each input C-place $P_i \in {}^\bullet T_j$. Then, the firing speed $v_j(t)$ of T_j is $v_j(t) = V_j(t)$.

Step2. T_j is called **weakly enabled** at time t if there exists a token changing from unavailable to available in each input C-place $P_i \in {}^\bullet T_j$ such that $M^a(P_i) = 0$ and $M^a(P_i) \geq I(P_i, T_j)$ for each input D-place $P_i \in {}^\bullet T_j$. Then, the firing speed $v_j(t)$ of T_j is

$$v_j(t) = \min\{V_j(t), \min_{P_i} \frac{pv_i(t)}{I(P_i, T_j)}\}$$

where $\displaystyle\min_{P_i}$ is taken for every input C-place $P_i \in {}^{\bullet}T_j$ such that $M^a(P_i) = 0$, and $pv_i(t)$ denotes the speed at which the unavailable tokens become available in P_i at time t.

Step3. T_j is called **firing-enabled** if it is strongly enabled or weakly enabled.

Step4. If T_j firing-enabled at time t continues to fire from the time t for δ, available tokens of $I(P_i,T_j) \times \int_{t}^{t+\delta} v_j(\tau)d\tau$ in each input C-place $P_i \in {}^{\bullet}T_j$ are taken away and unavailable tokens of $O(P_i,T_j) \times \int_{t}^{t+\delta} v_j(\tau)d\tau$ are added to each output C-place $P_i \in T_j^{\bullet}$.

Step5. The unavailable tokens added to each output C-place $P_i \in T_j^{\bullet}$ remain unavailable during the delay time of d_i. These unavailable tokens become available after d_i is passed.

We consider the following timed hybrid Petri net modeling a manufacturing system which produces 10 parts per one lot. The D-transition T_1 expresses importing 2 lots of parts which are not processed. The D-transition T_2 expresses exporting 2 lots of parts which are processed. The C-transition T_3 expresses processing parts where the maximum processing rate is V_3. The D-places P_1 and P_2 express the stop and execution of processing parts. The C-places P_4 and P_3 express parts which are processed and not processed. The D-place P_5 expresses the number of lots of imported parts. The initial marking is $M_0=(1,0,0,0,2)$.

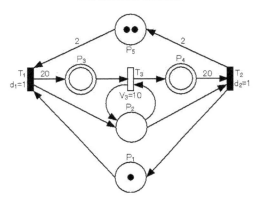

Figure 7.8 Timed Hybrid Petri Net of Manufacturing System

The behavior of the timed hybrid Petri net of the manufacturing system is as follows. From this behavior, it is found that the workings of the system is cyclic.

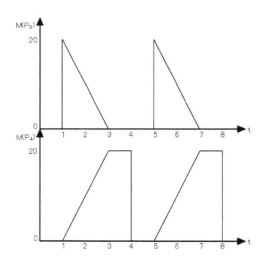

Figure 7.9 Behavior of Timed Hybrid Petri Net of Manufacturing System

Conflict Resolution

Hybrid Petri nets have six kinds of **structural conflicts** where more than one transition shares tokens in input places[6,7,8]. On the structural conflicts, the occurrence of a conflict is said to be an **effective conflict**.

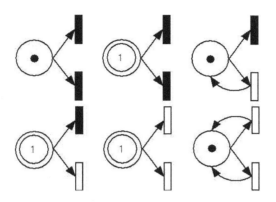

Figure 7.10 Structural Conflict

We consider a **conflict resolution** using a **linear programming approach**[12] in case where more than one C-transition shares tokens in input C-places. This approach is discussed in timed continuous Petri nets which are sub classes of timed hybrid Petri nets. However, it can be straightforwardly extended to timed hybrid Petri nets.

Let a set of input C-places without available tokens be P^{conf} for weakly enabled C-transitions. For each $P_i \in P^{conf}$, let a set of weakly enabled output C-transitions at time t be $T_{P_i}^{weak}$, and let a set of input C-transitions firing at time $t - d_i$ be $^{\bullet}P_i^{fire}$. Let a firing speed at time t be $v_j^{weak}(t)$ for each $T_j^{weak} \in T_{P_i}^{weak}$, and let a firing speed at time t be $v_{T_k^{fire}}(t)$ for each $T_k^{fire} \in {^{\bullet}P_i^{fire}}$. Thus the speed of tokens changing from unavailable to available in P_i at time t is expressed by

$$v_{P_i}^{in}(t) = \sum_{T_k^{fire} \in {^{\bullet}P_i^{fire}}} v_{T_k^{fire}}(t - d_i) \cdot O(P_i, T_k^{fire}).$$

The speed of available tokens taken away from P_i at time t is expressed by

$$v_{P_i}^{out}(t) = \sum_{T_j^{weak} \in T_{P_i}^{weak}} v_j^{weak}(t) \cdot I(P_i, T_j^{weak})$$
.

Also, the speed of available tokens taken away from P_i is less than or equal to that of tokens changing from unavailable to available in P_i at time t for each $P_i \in P^{conf}$, that is, $v_{P_i}^{in}(t) \geq v_{P_i}^{out}(t)$. In addition, firing speed is always non-negative and less than or equal to the maximum firing speed, then $V_j \geq v_j^{weak}(t) \geq 0$ for each $T_j^{weak} \in T_{P_i}^{weak}$.

Thus, let consider an optimization problem to maximize the firing speed $v_j^{weak}(t)$ for each $T_j^{weak} \in T_{P_i}^{weak}$ under such constraints. Then, let a priority coefficient of firing be α_j and we consider minimizing an objective function $V = \sum_j \alpha_j (V_j - v_j^{weak}(t))$. The bigger α_j is, the smaller $V_j - v_j^{weak}(t)$ is under the constraints for each $T_j^{weak} \in T_{P_i}^{weak}$ resulting in $v_j^{weak}(t)$ is bigger. Therefore, the optimization problem of $v_j^{weak}(t)$ for each $T_j^{weak} \in T_{P_i}^{weak}$ is formalized by a liner programming problem as follows:

$$\text{minimize } V = \sum_j \alpha_j (V_j - v_j^{weak}(t))$$

subject to $v_{P_i}^{in}(t) \geq v_{P_i}^{out}(t)$ and $V_j \geq v_j^{weak}(t) \geq 0$ $(\forall T_j^{weak} \in T_{P_i}^{weak}, \forall P_i \in P^{conf})$

For example, as shown in Fig.7.11, we consider modeling a tank system by a timed continuous Petri net. Liquid is provided into tank 1, 2 and 3 through valve1, 2 and 3, and taken away by pump 1 and 2 and valve 4.

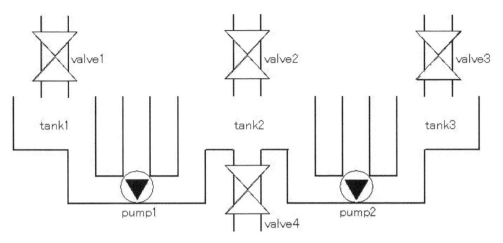

Figure 7.11 Tank System

As shown in Fig.7.12, the tank system is modeled by a timed continuous Petri net. The C-places P_1, P_2 and P_3 express the tank 1, 2 and 3. The C-transitions T_1, T_2, T_3 and T_4 express the valve 1, 2, 3 and 4. The C-transitions T_5 and T_6 express the pump 1 and 2.

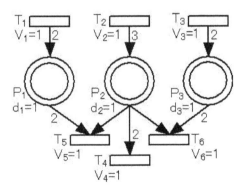

Figure 7.12 Continuous Petri Net of Tank System

The speeds of tokens changing from unavailable to available in each place P_i ($i = 1,2,3$) at time t are expressed as follows:

$$v_{P_1}^{in}(t) = O(P_1, T_1) \cdot v_1(t - d_1) = 2v_1(t-1)$$
$$v_{P_2}^{in}(t) = O(P_2, T_2) \cdot v_2(t - d_2) = 3v_2(t-1)$$
$$v_{P_3}^{in}(t) = O(P_3, T_3) \cdot v_3(t - d_3) = 2v_3(t-1).$$

The speeds of available tokens taken away from each place P_i ($i = 1,2,3$) at time t are expressed as follows:

$$v_{P_1}^{out}(t) = I(P_1, T_5) \cdot v_5(t) = 2v_5(t)$$

$$v_{P_2}^{out}(t) = I(P_2, T_5) \cdot v_5(t) + I(P_2, T_4) \cdot v_4(t) + I(P_2, T_6) \cdot v_6(t) = v_5(t) + 2v_4(t) + v_6(t)$$

$$v_{P_3}^{out}(t) = I(P_3, T_6) \cdot v_6(t) = 2v_6(t)$$

An effective conflict is occurred by C-transitions T_4, T_5 and T_6 for C-places P_1, P_2 and P_3 at time $t = 1$. Then, their optimal firing speeds $v_4(1)$, $v_5(1)$ and $v_6(1)$ are decided by resolving the conflict using linear programming approaches (a), (b) and (c) as follows:

In case of (a), let priority coefficients of firing for C-transitions T_4, T_5 and T_6 be $\alpha_4 = 5$, $\alpha_5 = 10$ and $\alpha_6 = 3$. Then, the minimized objective function is

$$V = \alpha_4(V_4 - v_4(1)) + \alpha_5(V_5 - v_5(1)) + \alpha_6(V_6 - v_6(1)) = 2.5.$$

Thus, the optimized firing speeds are

$$v_4(t) = 0.5, \ v_5(t) = 1 \text{ and } v_6(t) = 1.$$

Similarly, in case of (b), let $\alpha_4 = 10$, $\alpha_5 = 5$ and $\alpha_6 = 3$, then $V = 3$. Thus,

$$v_4(t) = 1, \ v_5(t) = 1 \text{ and } v_6(t) = 0.$$

In case of (c), let $\alpha_4 = 3$, $\alpha_5 = 5$ and $\alpha_6 = 10$, then $V = 1.5$. Thus,

$$v_4(t) = 0.5, \ v_5(t) = 1 \text{ and } v_6(t) = 1.$$

Therefore, the behaviors of the timed continuous Petri net of the tank system are shown as in Fig.7.13. From these behaviors, it is found that the optimal firing speeds are decided according to the priorities of firing and the conflicts are resolved. Note that in case of (b), the tokens only increase in P_3 because $v_6(t) = 0$.

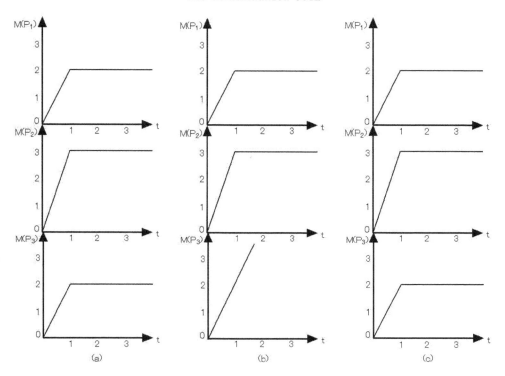

Figure 7.13 Behavior of Continuous Petri Net of Tank System

Chapter 8. Timed Event Graph

Sub Class of Petri Net

Petri nets include not only extended classes such as **colored Petri nets**[1], **high-level Petri nets**[2] and **hybrid Petri nets**[3,4,5], but also sub classes such as **state machines**[6,7], **event graphs**[7] and **free choice nets**[6]. There has been increasing numbers of net model classes, nowadays which forms the **net-oriented paradigm**[8]. This chapter addresses the performance evaluation of systems modeled by **timed event graphs**[9,10] and their relevant algebra.

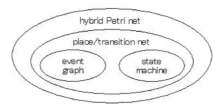

Figure 8.1 Net-oriented Paradigm

Timed Event Graph

Timed event graphs[9,10] are in a sub class with timed place/transition nets. Timed place/transition nets are timed discrete Petri nets[3,4,5] which are in a sub class with timed hybrid Petri nets[3,4,5]. Thus if R^+ is a set of positive real numbers, then

[Definition 8.1] Timed place/transition nets are defined by
$$TPN =< PN, Delay >$$

where PN is a place/transition net, and $Delay : P \rightarrow R^+ \cup \{0\}$ is a function assigning a non-negative real number to each place P_i, and $Delay(P_i) = d_i$ is a delay time of a token supplied to P_i by firing of the input transition until it can contribute to the firing of the output transition.

Event graphs are place/transition nets where the number of input or output transitions of a place is one, thus

[Definition 8.2] Timed event graphs are timed place/transition nets such that $|P_i^\bullet| = |{}^\bullet P_i| = 1$ ($\forall P_i \in P$).

Max-plus Algebra

The behavior of timed event graphs is expressed by the following max-plus algebra[9,10]. If R is a set of real numbers, then

[Definition 8.3] Max-plus algebra is an algebraic system consisting of a set $R \cup \{-\infty\}$ and two operators $\oplus$ and $\otimes$ defined on the set such that
$a \oplus b = \max(a,b)$ and $a \otimes b = a + b$ for $a, b \in R \cup \{-\infty\}$.

A zero or unit element for $\oplus$ or $\otimes$ is defined by $\varepsilon := -\infty$ or $e := 0$ respectively. The notation $\otimes$ is omitted below as the multiplication in general use.

In a timed event sub graph shown as in Fig.8.2, let the k-th firing time of the transition T_1 be $t_1(k)$ and the k-th arrival time of tokens in P_1 be $p_1(k)$, then it holds that $p_1(k) = t_1(k-2)$ ($k \geq 3$).

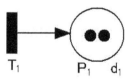

T_1 P_1 d_1

Figure 8.2 Simple Timed Event Sub Graph

In a timed event graph shown as in Fig.8.3, let the k-th firing time of the transition T_1 be $t_1(k)$ and the k-th arrival time of tokens in each $P_i \in {}^\bullet T_1$ be $p_i(k)$, then it holds that
$$t_1(k) = \max_{P_i \in {}^\bullet T_1}(p_i(k) + d_i).$$

The max-plus algebraic expression is written as

$$t_1(k) = \bigoplus_{P_i \in {}^\bullet T_1} p_i(k)d_i .$$

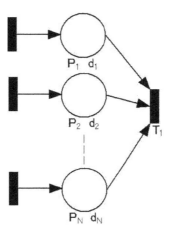

Figure 8.3 Simple Timed Event Graph

In a timed event graph shown as in Fig.8.4, let the k-th firing time of the transitions T_1, T_2 and T_3 be $t_1(k)$, $t_2(k)$ and $t_3(k)$, and the k-th arrival time of tokens in places P_1, P_2 and P_3 be $p_1(k)$, $p_2(k)$ and $p_3(k)$, then for $k \geq 2$ it holds that $p_1(k) = t_1(k-1)$, $p_2(k) = t_2(k-1)$ and $p_3(k) = t_3(k-1)$. Thus,

$$t_3(k) = \max\{p_1(k)+d_1, p_2(k)+d_2, p_3(k)+d_3\} = p_1(k)d_1 \oplus p_2(k)d_2 \oplus p_3(k)d_3$$

Therefore the arrival time of tokens in each place is expressed by

$$p_3(k) = p_1(k-1)d_1 \oplus p_2(k-1)d_2 \oplus p_3(k-1)d_3 \ (k \geq 2).$$

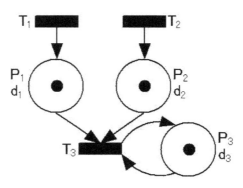

Dr. Atsushi Tanaka

Figure 8.4 Small Timed Event Graph

Transition Graph

In timed event graphs, each place has an input or output arc whose weight is one. So, if such a place and its input and output arcs are replaced by a **directed arc labeled by delay time**, and each transition is expressed by a **circle node**, then such graphs are said to be **transition graphs**[11].

Let us consider a timed event graph of a simple communication protocol[12] shown as in Fig.8.5. This graph is expressed by

$$\begin{bmatrix} p_1(k) \\ p_4(k) \end{bmatrix} = \begin{bmatrix} d_1 d_3 (d_2 \oplus d_5 d_7 d_8) & d_3 d_4 d_5 d_8 \\ d_1 d_5 d_6 d_7 & d_4 d_5 d_6 \end{bmatrix} \begin{bmatrix} p_1(k-1) \\ p_4(k-1) \end{bmatrix}$$

Let $d_i = 1\,(\forall i)$, then

$$\begin{bmatrix} p_1(k) \\ p_4(k) \end{bmatrix} = \begin{bmatrix} 5 & 4 \\ 4 & 3 \end{bmatrix} \begin{bmatrix} p_1(k-1) \\ p_4(k-1) \end{bmatrix}$$

Let a state vector be $X_k = [p_1(k), p_4(k)]^T$ with initial states $p_1(0) = p_4(0) = 0$, then X_k takes the following values:

$$X_1 = \begin{bmatrix} 5 \\ 4 \end{bmatrix},\ X_2 = \begin{bmatrix} 10 \\ 9 \end{bmatrix},\ X_3 = \begin{bmatrix} 15 \\ 14 \end{bmatrix},\ \text{and so on.}$$

Thus it is found that the minimum cycle is 5 time units.

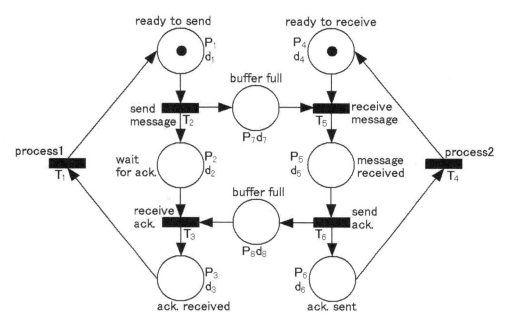

Figure 8.5 Timed Event Graph of Simple Communication Protocol

The timed event graph of Fig.8.5 is expressed by the transition graph shown as in Fig.8.6. Generally, in transition graphs, the mean weight of a directed circuit is calculated by the total delay time divided by the number of tokens in the directed circuit. Thus the minimum cycle of the communication protocol is obtained 5 time units from the **maximum mean weight of directed circuits** of Fig.8.6.

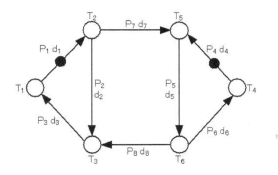

Figure 8.6 Transition Graph of Simple Communication Protocol

Dr. Atsushi Tanaka

Reduction of Timed Event Graph

In general, for analyzing dynamic properties of a large system, a **reduction method**[13] is used. This chapter discusses that a timed event graph is transformed into as simple graph as possible while preserving time behaviors based on max-plus algebra[14]. By this method, a performance evaluation index such as the **minimum cycle** or the **mean throughput** of a system is easily obtained, and structures preserving time behaviors are clarified.

For simplifying discussions, a timed event graph is expressed by a transition graph $TG =< N, Delay, M_0 >$ where N is a structure of TG consisting of a finite set of places P, a finite set of transitions T and their connections. $Delay : P \rightarrow R^+ \cup \{0\}$ is a labeling function to places and expresses a delay time associated with a place. A place is expressed by a labeled directed arc. A transition or token is an unfilled or filled circle respectively. M_0 is an initial marking.

First, we consider transformations of transition graphs shown as in Fig.8.7. Then a transition $t \in T$ is eliminated not to affect the k-th firing time $t_j(k)$ for each transition $t_j \in T$ ($j = 1,2,...,N$) except t. For example, the transition graph (c) of Fig.8.7 is expressed by

$$t_2(k) = d_2 t(k) \text{ and } t(k) = d_1 t_1(k-1) \oplus dt(k-1).$$

By eliminating $t(k)$,

$$t_2(k) = d_1 d_2 t_1(k-1) \oplus dt_2(k-1),$$

then the lower graph is obtained from the upper graph.

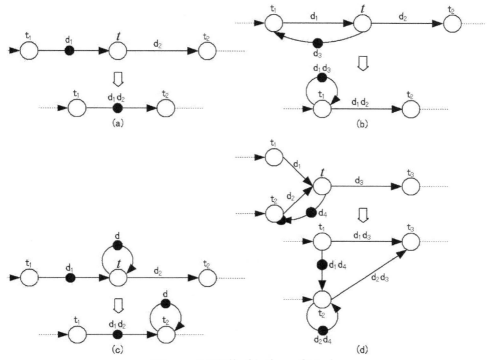

Figure 8.7 Elimination of Node

The generalized transformation of Fig.8.7 is Fig.8.8 where $^{\bullet}(^{\bullet}t)$ and $(t^{\bullet})^{\bullet}$ may have common transitions. After eliminating the transition t, a directed arc $t_{ik} \to t_{jl}$ from each $t_{ik} \in {}^{\bullet}(^{\bullet}t)$ to each $t_{jl} \in (t^{\bullet})^{\bullet}$ is inserted, and the delay time or the number of tokens for $t_{ik} \to t_{jl}$ is the sum of that for $t_{ik} \to t$ and $t \to t_{jl}$. If t has self-loop, the loop is added to each $t_{jl} \in (t^{\bullet})^{\bullet}$. Let Φ_{AT} denote this graph reduction method[15].

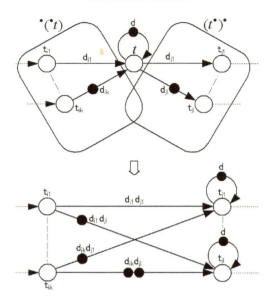

Figure 8.8 Reduction Method of Timed Event Graph

Φ_{AT} can be applied to any node of a strongly connected transition graph $TG = (N, Delay, M_0)$. The structure N includes a path from any node to any other node. Thus by eliminating any node, $TG' = (N', Delay', M_0')$ transformed from $TG = (N, Delay, M_0)$ is strongly connected as well.

Now, what relationships do hold in a directed circuit under Φ_{AT}? For example, we consider Fig.8.9. By eliminating the transition t of the upper transition graph $TG = (N, Delay, M_0)$, the lower transition graph $TG' = (N', Delay', M_0')$ is obtained. Then, the lower graph structure N' includes the directed circuit l_4 that the upper graph structure N does not include in addition to all the directed circuits l_1, l_2 and l_3 of N.

Note that l_4 is the direct sum of l_1 and l_2 in terms with arcs, delay time and the number of tokens. Thus if TG is live, TG' is live as well. Moreover, the mean weight of a directed circuit l in timed event graphs is defined by the total delay time in l divided by the sum of tokens in l, then for TG and TG',

the mean weight of l_4

$$= \frac{d'_2 + d'_3 + d'_4 + d'_5}{2} \leq \max(d'_2 + d'_3, d'_4 + d'_5) = \max(\text{the weight of } l_1, \text{ the weight of } l_2)$$

It follows that the maximum mean weight of a directed circuit of TG is the same as that of TG'. Therefore, the maximum mean weight of a directed circuit is invariable under Φ_{AT}.

The properties found in Fig.8.9 hold in general, that is, when a transition t is eliminated by Φ_{AT}, a generated directed circuit is the direct sum of directed circuits including t in terms with arcs, delay time and tokens.

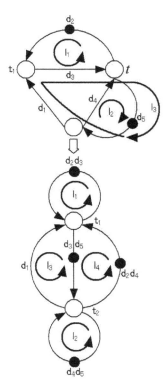

Figure 8.9 Generated Directed Circuit by Elimination of Node

Φ_{AT} is repetitively applied to a transition graph $TG = (N, Delay, M_0)$ while a transition is continued eliminating, finally a transition graph

93

Dr. Atsushi Tanaka

$TG_F = (N_F, Delay_F, M_F)$ consisting of a single transition t_F and its self-loops shown as in Fig.8.10 is obtained. The mean weight of each self-loop implies the delay time for firing of t_F. Obviously, the minimum cycle of TG_F is equal to the maximum mean weight of self-loops.

The minimum cycle of TG_F is equal to that of TG. Also, the maximum mean weight of TG_F is the same as that of TG under Φ_{AT}. After all, it follows from the minimum cycle of a transition graph $TG = (N, Delay, M_0)$ that the minimum cycle of $(N, Delay, M_0)$ is equal to the maximum mean weight of directed circuits of $(N, Delay, M_0)$.

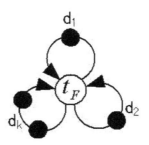

Figure 8.10 Single Node Transition Graph

As discussed above, the following property holds for strongly connected timed event graphs.

[Property 8.1] **Strong connectivity**, **liveness** and **the maximum mean weight** of directed circuits are invariable under Φ_{AT} applied to strongly connected timed event graphs.

Note that the **mean throughput** of a system is given by the reciprocal of the minimum cycle, hence the throughput is also invariable under Φ_{AT}.

Minimum Cycle of Timed Event Graph

We consider the minimum cycle of a typical net structure for a manufacturing system shown as in Fig.8.11[14]. An unit sub net of a timed event graph is expressed by (a) and a timed event graph chained N unit sub nets is expressed by (b).

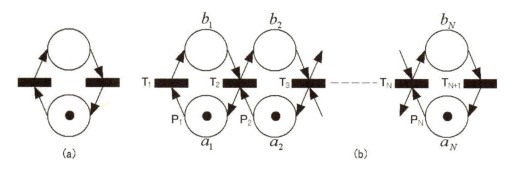

Figure 8.11 Chained Unit Sub Net

Let the k-th arrival time of tokens for each place P_i (i=1,...,N) with an initial token be $p_i(k)$. Then the following equation is obtained:

$$\begin{bmatrix} p_1(k) \\ p_2(k) \\ \vdots \\ \vdots \\ \vdots \\ p_N(k) \end{bmatrix} = \begin{bmatrix} a_1 b_1 & a_2 & \varepsilon & \cdots & \cdots & \varepsilon \\ a_1 b_1 b_2 & a_2 b_2 & a_3 & \ddots & & \vdots \\ \vdots & \vdots & a_3 b_3 & \ddots & \ddots & \vdots \\ \vdots & \vdots & \vdots & \ddots & \ddots & \varepsilon \\ \vdots & \vdots & \vdots & & \ddots & a_N \\ a_1 b_1 \cdots b_N & a_2 b_2 \cdots b_N & a_3 b_3 \cdots b_N & \cdots & \cdots & a_N b_N \end{bmatrix} \begin{bmatrix} p_1(k-1) \\ p_2(k-1) \\ \vdots \\ \vdots \\ \vdots \\ p_N(k-1) \end{bmatrix}$$

Let the coefficient matrix of this equation be **M**. The graph G(**M**) consisting of N nodes shown as in Fig.8.12 where the weight of a directed arc from the i-th node to the j-th node is given by the (i, j) element of **M** is obtained. Then the mean weight of a directed circuit in G(**M**) is considered where the mean weight is given by the weight divided by the length in a directed circuit.

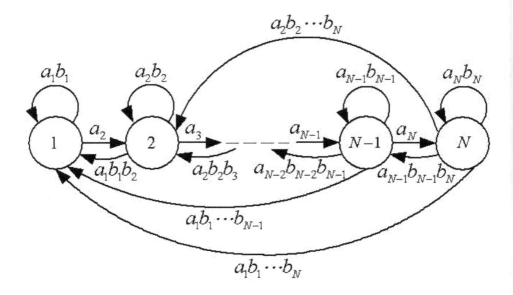

Figure 8.12 Graph G(**M**) of Coefficient Matrix **M**

N directed circuits are generated by the addition of node N, and their length and weight are shown as in Fig.8.13. Then if $a_N + b_N \geq a_i + b_i$ ($i \neq N$), the maximum mean weight of directed circuits generated by the additional node N is $a_N + b_N$. Thus the maximum mean weight of G(**M**) is $a_N + b_N$ which implies the mean weight of the N-th unit sub net. Also, the same discussion holds in case where the number of nodes in G(**M**) is i ($i \neq N$). Then the maximum mean weight of G(**M**) is given by the maximum value of the mean weights of unit sub nets $\max(a_1 + b_1, a_2 + b_2, \cdots, a_N + b_N)$. Therefore the minimum cycle of the total system is the maximum value of the minimum cycles in unit sub nets.

length	weight
1	$a_N + b_N$
2	$(a_N + b_N) + (a_{N-1} + b_{N-1})$
$\vdots$	$\vdots$
N	$(a_N + b_N) + (a_{N-1} + b_{N-1}) + \cdots + (a_1 + b_1)$

Figure 8.13 Generated Directed Circuit by Addition of Node N

Next, by applying Φ_{AT} to the timed event graph (b) of Fig.8.11, the minimum cycle is obtained. The graph (b) of Fig.8.11 is expressed by the upper transition graph of Fig.8.14. By applying Φ_{AT} to the transition graph and eliminating nodes T_1, T_2, ..., T_N, the lower transition graph of Fig.8.14 is obtained.

Then, the maximum mean weight is the maximum value of the mean weights in the unit sub nets. Thus the minimum cycle of the total system is given by the maximum mean weight $\max(a_1 + b_1, a_2 + b_2, \cdots, a_N + b_N)$.

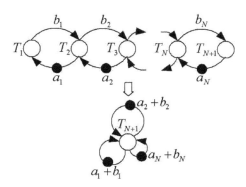

Figure 8.14 Reduction of Chained Transition Graph

Mean Throughput of Manufacturing System

We consider the mean throughput of a flow shop manufacturing system shown as in Fig.8.15[9]. A flow shop is a manufacturing line where parts are processed through several machines and similar products can be manufactured.

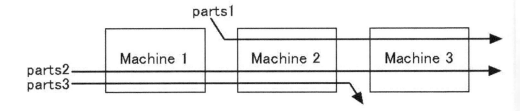

Figure 8.15 Flow Shop Manufacturing System

In this manufacturing system, three kinds of parts 1, 2 and 3 are processed through three machines 1, 2 and 3 in series and different products peculiar to each machine are manufactured. The processing time of parts by each machine is shown as in Fig.8.16[9]. And this manufacturing system is expressed by the timed event graph of Fig.8.17[14]. The node semantics of the graph is shown as in Fig.8.18.

	parts1	parts2	parts3
Machine 1	–	1	5
Machine 2	3	2	3
Machine 3	4	3	–

Figure 8.16 Processing Time of Parts by Machine

The timed event graph of Fig.8.17 is expressed by the transition graph of Fig.8.19. By eliminating transitions T_2 and T_6 of Fig.8.19, the reduced transition graph of Fig.8.20 is obtained. By eliminating transitions T_4 of Fig.8.20, the reduced transition graph of Fig.8.21 is obtained. Finally, by eliminating transitions

T_3, T_5 and T_7 of Fig.8.21, the reduced single node transition graph of Fig.8.22 is obtained. The maximum mean weight of the transition graph of Fig.8.22 is given by

$$\max(23/4, 22/3, 19/2, 8/1) = 19/2 \,.$$

Therefore the mean throughput of the manufacturing system is the reciprocal of the maximum mean weight $2/19$ parts/time unit.

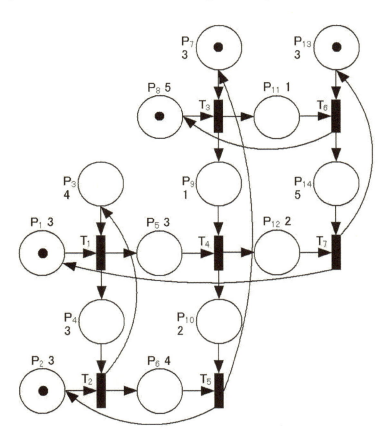

Figure 8.17 Timed Event Graph of Manufacturing System

node	semantics
P_1	Machine 2 is empty
P_2	Machine 3 is empty
P_3	Processing parts1 is completed
P_4	Transferring parts1 from Machine 2 to Machine 3
P_5	Processing parts1 by Machine 2
P_6	Processing parts1 by Machine 3
P_7	Processing parts2 is completed
P_8	Machine 1 is empty
P_9	Transferring parts2 from Machine 1 to Machine 2
P_{10}	Transferring parts2 from Machine 2 to Machine 3
P_{11}	Processing parts2 by Machine 1
P_{12}	Processing parts2 by Machine 2
P_{13}	Processing parts3 is completed
P_{14}	Transferring parts3 from Machine 1 to Machine 2
T_1	Start processing parts1 by Machine 2
T_2	Start processing parts1 by Machine 3
T_3	Start processing parts2 by Machine 1
T_4	Start processing parts2 by Machine 2
T_5	Start processing parts2 by Machine 3
T_6	Start processing parts3 by Machine 1
T_7	Start processing parts3 by Machine 2

Figure 8.18 Node Semantics of Timed Event Graph of Manufacturing System

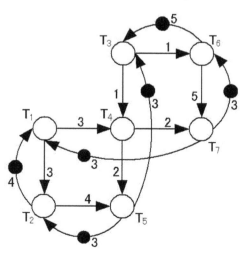

Figure 8.19 Transition Graph of Timed Event Graph of Manufacturing System

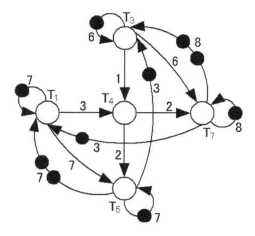

Figure 8.20 Reduced Transition Graph from Fig.8.19

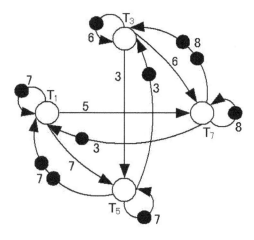

Figure 8.21 Reduced Transition Graph from Fig.8.20

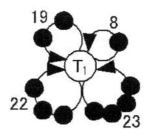

Figure 8.22 Reduced Single Node Transition Graph from Fig.8.21

Min-plus Algebra

Min-plus algebra[16] has the same algebraic structure as max-plus algebra. If R is a set of real numbers, then

[Definition 8.4] Min-plus algebra is an algebraic system consisting of a set $R \cup \{+\infty\}$ and two operators $\oplus$ and $\otimes$ defined on the set such that

$a \oplus b = \min(a,b)$ and $a \otimes b = a+b$ for $a,b \in R \cup \{+\infty\}$

A zero or unit element for $\oplus$ or $\otimes$ is defined by $\varepsilon := +\infty$ or $e := 0$ respectively. The notation $\otimes$ is omitted below as the multiplication in general use.

Fluid timed event graphs where the number of tokens is extended from non-negative integers to non-negative real numbers have been proposed[16]. The behavior of fluid timed event graphs is expressed by min-plus algebra. The following discussion holds for timed event graphs as well as fluid timed event graphs.

In fluid timed event graphs, let a set of input places for each transition $t_j \in T$ be P^j, delay time associated with each place $p_i \in P^j$ be d_i and an input transition of p_i be t_i. Then, the counts of firing of t_i till the time k is expressed by

$$t_j(k) = \min_{p_i \in P^j}(t_i(k-d_i) + M_0(p_i)).$$

The min-plus algebraic expression is written as

$$t_j(k) = \bigoplus_{p_i \in P^j} M_0(p_i) t_i(k - d_i).$$

Obviously, Φ_{AT} holds in fluid timed event graphs as well. For example, the upper transition graph (c) of Fig.8.7 is written as the following min-plus expressions:

$$t_2(k) = t(k - d_2) \text{ and } t(k) = t_1(k - d_1) \oplus t(k - d).$$

By eliminating $t(k)$,

$$t_2(k) = t_1(k - d_1 - d_2) \oplus t_2(k - d).$$

Then, the lower graph is obtained from the upper graph.

Note that the mean weight of a directed circuit is defined by the total number of tokens divided by the total delay time in the directed circuit of a fluid timed event graph. Also, the minimum cycle of a fluid timed event graph is the reciprocal of the minimum mean weight of the directed circuits. Therefore, in fluid timed event graphs as well as timed event graphs, the mean throughput or minimum cycle of the total system is easily obtained by Φ_{AT}.

Performance Evaluation of Timed Hybrid Petri Net

If the sub system which expresses the machine 1 in Fig.8.17 is modeled by a fluid timed event graph, the timed hybrid Petri net of Fig.8.23 where T_3 and T_6 are expressed by C-transitions and P_7, P_8, P_9, P_{11}, P_{13} and P_{14} are expressed by C-places is obtained[17].

By applying Φ_{AT} to the hybrid system for eliminating transitions except T_1, the single node transition graph of Fig.8.22 is obtained in the same way as applying to the timed event graph of Fig.8.17. Therefore, the mean throughput $2/19$ parts/time unit is easily obtained from the minimum mean weight of the graph.

Dr. Atsushi Tanaka

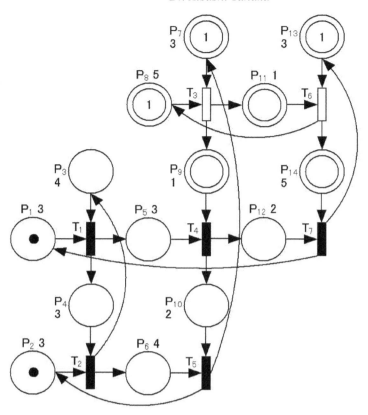

Figure 8.23 Timed Hybrid Petri Net of Flow Shop Manufacturing System

104

Chapter 9. System Architecture

System Framework

Designing systems is based on the architecture which is a closed-loop consisting of a **supervisor** part and a **plant** part called the **Ramadge-Wonham framework**, or the **RW framework** for short, as shown in Fig.9.1[1,2].

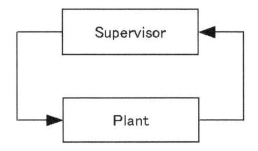

Figure 9.1 RW Framework

A subsystem which directly collaborates with a user should be divided as a **presentation** part shown as in Fig.9.2. A presentation part is indirectly connected with a plant part through a supervisor part. This architecture is said to be the **Presentation-Supervisor-Plant** architecture, or for short the **PSP architecture**.

```
        ┌──────────────────────┐
    ┌──▶│     Presentation     │──┐
    │    └──────────────────────┘  │
    │                              │
    │    ┌──────────────────────┐  │
    │ ┌─▶│     Supervisor       │◀─┘
    │ │   └──────────────────────┘◀─┐
    │ │                            │
    │ │   ┌──────────────────────┐ │
    └──┼─▶│        Plant         │─┘
      └──└──────────────────────┘
```

Figure 9.2 PSP Architecture

To classify model elements based on the PSP architecture, notations stereo type **<<Presentation>>**, **<<Supervisor>>** and **<<Plant>>** are used.

For example, in the class diagram shown as in Fig.9.3, the class Item Window is a presentation class collaborating with a user. The class Item Control is a supervisor class, controlling a presentation or plant class. The classes Item and Item List are plant classes holding and processing a data.

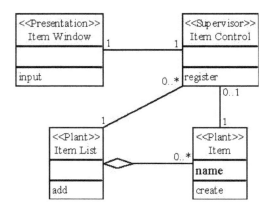

Figure 9.3 Class Diagram

An evolution of this system is expressed by the sequence diagram shown as in Fig.10.4. In this context, a clerk inputs an item data on a terminal window at first. Then the Item Window sends a message to the Item Control. The Item Control creates an object of the class Item in order to hold the item data and sends a message to the Item List in order to hold the created object.

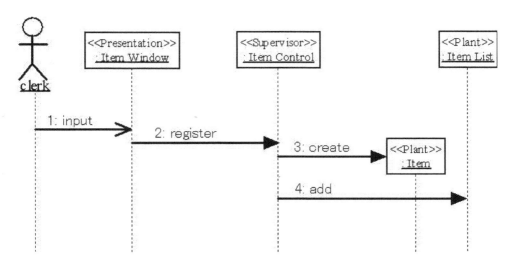

Figure 9.4 Sequence Diagram

Theoretical Foundation

Designing the framework is quite significant. However, the optimal design of complex systems is not easy in general. Although Petri nets are useful to model dynamics of complex systems, a heuristically modeled Petri net often becomes an **over-specification**. It will include a redundant specification. Thus we consider **controlled timed hybrid Petri nets with external input D-places**[3] as an extended class of Petri nets for the purpose of the framework design. Then necessary and sufficient conditions for the optimal design architecture are presented. If Z^+ is a set of positive integers and R^+ is a set of positive real numbers, then

[Definition 9.1] Controlled hybrid Petri nets with external input D-places are defined by

$$H^{ctl} = < P \cup P_{cp}, T, I \cup I_{cp}, O, h, M_0 > \quad (1)$$

where P is a finite set of internal places, T is a finite set of transitions, $P \cap T = \phi$ and $P \cup T \neq \phi$. $h : P \cup T \rightarrow \{D, C\}$ is a hybrid function that indicates whether each node is associated with a discrete variable or a continuous variable. $I : P \times T \rightarrow R^+ \cup \{0\}$ denotes a weight of an input arc of a transition. If $h(P_i) = D$, a constraint $I(P_i, T_j) \in Z^+ \cup \{0\}$ is added. $O : P \times T \rightarrow R^+ \cup \{0\}$ denotes a weight of an output arc of a transition. If $h(P_i) = D$, a constraint $O(P_i, T_j) \in Z^+ \cup \{0\}$ is added. When $h(P_i) = D \wedge h(T_j) = C$, an arc connecting P_i and T_j is said to be a permission arc such that $I(P_i, T_j) = O(P_i, T_j)$. P_{cp} is a finite set of external input D-places. $I_{cp} : P_{cp} \times T \rightarrow \{0,1\}$ is a function associating a permission arc from an external input D-place to a transition with the weight such that $I_{cp}(P_{cp_i}, T_j) = 1$. M_0 is an initial marking which is a non-negative integer for an internal D-place, or a non-negative real number for an internal C-place.

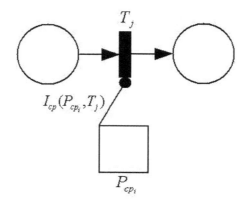

Figure 9.5 Permission Arc with External Input D-Place

[Definition 9.2] Controlled Timed hybrid Petri nets with external input D-places are defined by

$$H^{tim} =< H^{ctl}, Tempo, Delay > \quad (2)$$

where H^{ctl} is a controlled model defined by the expression (1). $Tempo : T \times R^+ \to R^+$ is a function assigning a positive real number to each C-transition T_j, and $Tempo(T_j, t) = V_j(t)$ indicates the maximum instantaneous firing speed of T_j at time t. $Delay : P \to R^+ \cup \{0\}$ is a function assigning a non-negative real number to each internal place P_i, and $Delay(P_i) = d_i$ is a delay time of a token supplied to P_i by firing of the input transition until it can contribute to firing of the output transition.

On controlled timed models defined by the expression (2), $M : P \to R^+ \cup \{0\}$ is a marking function over internal places where $M(P_i) \in Z^+ \cup \{0\}$ when $h(P_i) = D$. Also, $M_{cp} : P_{cp} \to \{0,1\}$ is a marking function over external input D-places determined by state feedback which will be discussed later. The above marking functions are equivalent to markings themselves, thus the notations M and M_{cp} are regarded as internal and external markings, respectively.

[Definition 9.3] Firing rules of a controlled D-transition are defined as follows:

Step1. A D-transition T_j is called **marking-enabled** at time t if it is a source transition or $M^a(P_i) \geq I(P_i, T_j)$ for each internal input place $P_i \in {}^\bullet T_j$.

Step2. A D-transition T_j is called **control-enabled** at time t if $M_{cp}(P_{cp_i}) \geq I_{cp}(P_{cp_i}, T_j)$ for each external input D-place $P_{cp_i} \in {}^\bullet T_j$.

Step3. T_j is called **firing-enabled** if it is marking-enabled and control-enabled.

Step4. If T_j firing-enabled at time t fires, $I(P_i, T_j)$ available tokens are taken away from each internal input place $P_i \in {}^\bullet T_j$ and $O(P_i, T_j)$ unavailable tokens are added to each internal output place $P_i \in T_j^\bullet$.

Step5. The unavailable tokens added to each internal output place $P_i \in T_j^\bullet$ remain unavailable during the delay time of d_i. These unavailable tokens become available after d_i is passed.

[Definition 9.4] Firing rules of a controlled C-transition are defined as follows:

Step1. A C-transition T_j is called **strongly enabled** at time t if it is a source transition or $M^a(P_i) \geq I(P_i, T_j)$ for each internal input D-place $P_i \in {}^\bullet T_j$ and $M^a(P_i) > 0$ for each internal input C-place $P_i \in {}^\bullet T_j$. Then, the firing speed $v_j(t)$ of T_j is $v_j(t) = V_j(t)$.

Step2. T_j is called **weakly enabled** at time t if there exists a token changing from unavailable to available in each internal input C-place $P_i \in {}^\bullet T_j$ such that $M^a(P_i) = 0$ and $M^a(P_i) \geq I(P_i, T_j)$ for each internal input D-place $P_i \in {}^\bullet T_j$. Then, the firing speed $v_j(t)$ of T_j is

$$v_j(t) = \min\{V_j(t), \min_{P_i} \frac{pv_i(t)}{I(P_i, T_j)}\}$$

where $\min_{P_i}$ is taken for every internal input C-place $P_i \in {}^\bullet T_j$ such that $M^a(P_i) = 0$, and $pv_i(t)$ denotes the speed at which the unavailable tokens become available in P_i at time t.

Step3. T_j is called **marking-enabled** if it is strongly enabled or weakly enabled.

Step4. A C-transition T_j is called **control-enabled** at time t if $M_{cp}(P_{cp_i}) \geq I_{cp}(P_{cp_i}, T_j)$ for each external input D-place $P_{cp_i} \in T_j$.

Step5. T_j is called **firing-enabled** if it is marking-enabled and control-enabled.

Step6. If T_j firing-enabled at time t continues to fire from the time t for δ, available tokens of $I(P_i, T_j) \times \int_t^{t+\delta} v_j(\tau)d\tau$ in each internal input C-place $P_i \in {}^\bullet T_j$ are taken away and unavailable tokens of $O(P_i, T_j) \times \int_t^{t+\delta} v_j(\tau)d\tau$ are added to each internal output C-place $P_i \in T_j^\bullet$.

Step7. The unavailable tokens added to each internal output C-place $P_i \in T_j^\bullet$ remain unavailable during the delay time of d_i. These unavailable tokens become available after d_i is passed.

Let $R(H^{tim})$ denote the closure of a reachable set for H^{tim}. And let $M[T_j >$ and $M[T_j > M'$ denote that $T_j \in T$ is marking-enabled at $M \in R(H^{tim})$, and that M' is reachable from M through the firing of T_j, respectively. It is addressed that two or more transitions fire concurrently, thus 2^T is used to denote a power set of T. Let $M[T^{sub} >$ and $M[T^{sub} > M'$ denote that $T^{sub} \in 2^T$ is marking-enabled at $M \in R(H^{tim})$, and that M' is reachable from M through the concurrent firing of T^{sub}, respectively. A subset $T_b \subseteq 2^T$ is defined by

$$T_b = \{T_{b_j} \in 2^T \mid M[T_{b_j} > \text{ for some } M \in R(H^{tim})\}$$

Each $T_{b_j} \in T_b$ is divided so that $T_{b_j} = {}^D T_{b_j} \cup {}^C T_{b_j}$ where ${}^D T_{b_j}$ and ${}^C T_{b_j}$ are sets of D- and C-transitions, respectively.

Let $B = \{0,1\}^{R(H^{tim})}$ be a set of predicates on $R(H^{tim})$. It is said that a predicate $Q \in B$ is true (respectively, false) at $M \in R(H^{tim})$ if $Q(M) = 1$ (respectively, =0), and the only closed sets of markings at which a predicate is true are considered. Let $\neg$, $\wedge$ and $\vee$ denote the fundamental Boolean two terms operators negation, conjunction and disjunction on B, respectively. So that readers do not confuse a predicate with a set of markings, let $\mathcal{Q}$ denote a set of markings at which $\mathcal{Q}$ is true. A partial order $\leq$ on B is defined as follows: For arbitrary $Q_1, Q_2 \in B$, $Q_1 \leq Q_2$ iff $Q_1(M) \leq Q_2(M)$ at every $M \in R(H^{tim})$.

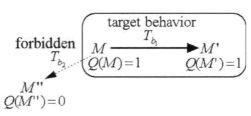

Figure 9.6 Control Specification

A predicate $D_{T_{b_j}}$ implying whether more than one transition T_{b_j} is marking-enabled or not is defined as follows: There exists $M \in R(H^{tim})$, for all $T_{b_j} \in T_b$,

$$D_{T_{b_j}}(M) = \begin{cases} 1 \text{ if } M[T_{b_j} > \\ 0 \quad \text{otherwise} \end{cases}$$

A predicate transformer[4] $wp_{T_{b_j}}$ on B for more than one transition implying whether a reachable marking after its firing satisfies $Q \in B$ or not is defined at $M \in R(H^{tim})$ at time t:

case(i) $T_{b_j} = {}^D T_{b_j}$: if $M[T_{b_j} > M'$ and $Q(M') = 1$, then $wp_{T_{b_j}}(Q)(M) = 1$;

case(ii) $T_{b_j} = {}^C T_{b_j}$: if T_{b_j} fires continuously, there exists a small enough $\delta > 0$ and $Q(M') = 1$ at a marking M' at time $t + dt$ for $0 < dt \le \delta$, then $wp_{T_{b_j}}(Q)(M) = 1$;

case(iii) $T_{b_j} = {}^D T_{b_j} \cup {}^C T_{b_j}$: if $M[{}^D T_{b_j} > M'$, $Q(M') = 1$ and $Q(M'') = 1$ at $M'' \in R(H^{tim})$ reachable from M' through continuous firing of ${}^C T_{b_j}$, then $wp_{T_{b_j}}(Q)(M) = 1$;

case(iv) otherwise: $wp_{T_{b_j}}(Q)(M) = 0$.

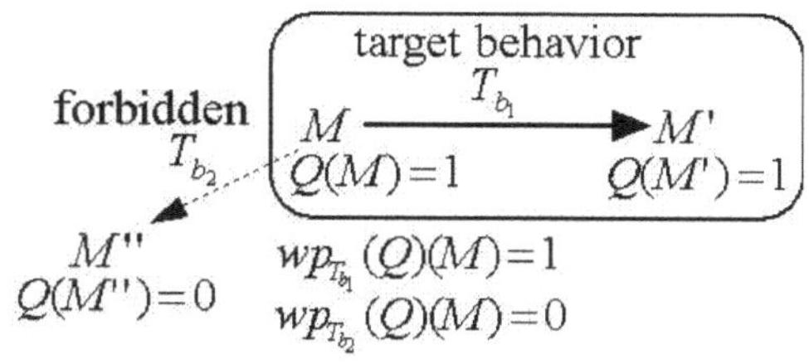

Figure 9.7 Predicate Transformer

In addition, a predicate transformer $wlp_{T_{b_j}}$ on B for each T_{b_j} is defined by

$$wlp_{T_{b_j}}(Q) = wp_{T_{b_j}}(Q) \vee \neg D_{T_{b_j}} \quad (3)$$

The expression (3) implies that $wlp_{T_{b_j}}(Q)(M)$ is a transformer that is 1 for transition to a state satisfying Q by simultaneous firing of T_{b_j} or concurrently firing-disabled transitions T_{b_j} for the marking M, and is 0 otherwise.

State Feedback

Two kinds of sets of controllable and uncontrollable transitions T_c and T_u are defined by

$$T_c = \{T_j \in T \mid I_{cp}(P_{cp_i}, T_j) = 1 \text{ for some } P_{cp_i} \in P_{cp}\},$$
$$T_u = T \setminus T_c = \{T_j \in T \mid T_j \notin T_c\}.$$

A subset of external input D-places $^{cp}T_{b_j} \subseteq P_{cp}$ for each $T_{b_j} \in T_b$ is defined by

$$^{cp}T_{b_j} = \{P_{cp_i} \in P_{cp} \mid I_{cp}(P_{cp_i}, T_j) = 1 \text{ for some } T_j \in T_{b_j}\}.$$

Let Γ denote a set of control patterns given by a power set of P_{cp}. A control pattern $\gamma \in \Gamma$ is namely a set of external input D-places in which the existence of a token is determined by state feedback. The state feedback is defined as mapping from each $M \in R(H^{tim})$ to some $\gamma \in \Gamma$. Let $\Gamma^{R(H^{tim})}$ denote a set of state feedbacks, and let $H^{tim} \mid f$ denote H^{tim} with $f \in \Gamma^{R(H^{tim})}$. Obviously, a reachable set of a closed loop $R(H^{tim} \mid f)$ depends on the state feedback f. Thus a partial order $\leq$ on $\Gamma^{R(H^{tim})}$ is defined as follows: For arbitrary $f_1, f_2 \in \Gamma^{R(H^{tim})}$, $f_1 \leq f_2$ iff $f_1(M) \subseteq f_2(M)$ at every $M \in R(H^{tim})$.

For arbitrary $f_1, f_2 \in \Gamma^{R(H^{tim})}$, the sum of $f_1 + f_2$ is defined as follows: For each $M \in R(H^{tim})$,

$$(f_1 + f_2)(M) = f_1(M) \cup f_2(M).$$

A marking over external input D-places M_{cp} at time t is given by a state feedback $f \in \Gamma^{R(H^{tim})}$, thus there exists $M \in R(H^{tim})$, for all $P_{cp_i} \in P_{cp}$,

$$M_{cp}(P_{cp_t})(M)(t) = \begin{cases} 1 \text{ if } P_{cp_t} \in f(M) \\ 0 \qquad \text{otherwise} \end{cases}$$

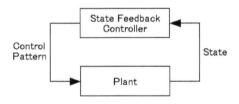

Figure 9.8 Closed Loop with State Feedback

Control Invariance

For any $f \in \Gamma^{R(H^{tim})}$, a predicate $f_{T_{b_j}} \in B$ is defined as follows: There exists $M \in R(H^{tim})$ and for all $T_{b_j} \in T_b$,

$$f_{T_{b_j}}(M) = \begin{cases} 1 \text{ if } {}^{cp}T_{b_j} \subseteq f(M) \\ 0 \qquad \text{otherwise} \end{cases}$$

If a state feedback $f \in \Gamma^{R(H^{tim})}$ satisfies the following expression for any $T_{b_j} \in T_b$, f is said to be a **permissive feedback**, or for short a **PF**, for Q:

$$Q \leq wlp_{T_{b_j}}(Q) \vee \neg f_{T_{b_j}} \quad (4)$$

The equation (4) implies that all markings reachable from an arbitrary marking satisfying Q in $H^{tim} | f$ satisfy Q.

When there exists a PF for Q, Q is said to be **control-invariant**. If Q is control-invariant, there is in general more than one PF. Let a set of all permissive feedbacks for Q be $F(Q)$.

A state feedback $\hat{f} \in F(Q)$ is said to be a **maximally permissive feedback**, or for short a **maximally PF**, for Q if no feedback $f (\neq \hat{f}) \in F(Q)$ satisfies $f \geq \hat{f}$. Generally there exists more than one maximally PF.

[Lemma 9.1] When a predicate Q is control-invariant, a maximally PF always exists in $F(Q)$.

(Proof) For arbitrary $f \in F(Q)$, let $R_f = R(H^{tim} | f)$. Obviously, R_f is a closed subset of Q. If $R(Q) = \{R_f \mid f \in F(Q)\}$, then there exists a maximal element $R_{f^*} (f^* \in F(Q))$ in $R(Q)$ because Q is a closed set. Further, let $\hat{F}(f^*) = \{f \in F(Q) \mid R_{f^*} = R_f\}$. A maximal element $\hat{f}^*$ always exists in $\hat{F}(f^*)$, because the number of tokens that can be inserted into an external input D-place is at most 1. If there exists f satisfying $f \geq \hat{f}^*$, then $R_f = R_{\hat{f}^*}$ and $f = \hat{f}^*$ from the definition of $\hat{f}^*$. Hence, there always exists a maximally PF in $F(Q)$. Q.E.D.

Optimal Design Architecture

If a maximally PF exists uniquely, it is said to be the **maximum permissive feedback**, or for short **MPF**. In general, a MPF does not always exist.

[Lemma 9.2] When a predicate Q is control-invariant, the MPF exists for Q iff Q satisfies the following condition (C).

(C) Arbitrary $f, g \in F(Q)$ satisfy the following expression for an arbitrary $T_{b_j} \in T_b$:

$$Q \leq wlp_{T_{b_j}}(Q) \vee f_{T_{b_j}} \vee g_{T_{b_j}} \vee \neg(f+g)_{T_{b_j}} \qquad (5)$$

(Proof) First, sufficiency is proven. From Lemma 9.1, a maximally PF always exists. It is now assumed that two maximally PF f_1 and f_2 exist such that $f_1 \neq f_2$. If $f_m = f_1 + f_2$, then f_m is clearly not a PF. Hence, there exist $T_{b_j} \in T_b$ and $M \in R(H^{tim})$ such that $Q(M) = 1$ and $wlp_{T_{b_j}}(Q)(M) \vee \neg f_{m_{T_{b_j}}}(M) = 0$. Then, $wlp_{T_{b_j}}(Q)(M) = 0$. Also, since $Q \leq wlp_{T_{b_j}}(Q) \vee \neg f_{1_{T_{b_j}}}$ holds from the fact that $f_1 \in F(Q)$, $f_{1_{T_{b_j}}}(M) = 0$. Similarly for $f_2 \in F(Q)$, $f_{2_{T_{b_j}}}(M) = 0$. Therefore, by virtue of the equation (5), $\neg(f_1+f_2)_{T_{b_j}} = 1$. This results in $f_{m_{T_{b_j}}}(M) = (f_1+f_2)_{T_{b_j}}(M) = 0$. Finally, $wlp_{T_{b_j}}(Q)(M) \vee \neg f_{m_{T_{b_j}}}(M) = 1$, which is a contradiction.

Next, necessity is proven. Obviously, the equation (5) holds for $M \in R(H^{tim})$ satisfying $Q(M) = 0$ or $wlp_{T_{b_j}}(Q)(M) = 1$. Hence, in what follows, let us consider $M \in R(H^{tim})$ satisfying $Q(M) = 1$ and $wlp_{T_{b_j}}(Q)(M) = 0$. Let the MPF for Q be f_m. Then, from the definition of the PF, $wlp_{T_{b_j}}(Q)(M) \vee \neg f_{m_{T_{b_j}}}(M) = 1$. Since $wlp_{T_{b_j}}(Q)(M) = 0$, it is found that $f_{m_{T_{b_j}}}(M) = 0$. On the other hand, from the definition of the MPF and that of the sum of state feedbacks, $f_m \geq f + g$ for arbitrary $f, g \in F(Q)$. Hence, since $f_{m_{T_{b_j}}}(M) = 0$, $(f+g)_{T_{b_j}}(M) = 0$ is obtained. From this, the equation (5) holds. Q.E.D.

From the proof for Lemma 9.2, the following corollary can easily be proven:

[Corollary 9.1] If the condition (C) in Lemma 9.2 holds, $f + g \in F(Q)$ for arbitrary $f, g \in F(Q)$.

Let $^{cp}T_j$ be a set of external input D-places connected to each controllable transition $T_j \in T_c$. Then, a set of transitions $T(P_{cp_i}) \subseteq T_c$ is defined as follows for each $P_{cp_i} \in P_{cp}$:

$$T(P_{cp_i}) = \{T_j \in T_c | {}^{cp}T_j = \{P_{cp_i}\}\}$$

Therefore, $T(P_{cp_i})$ is a set of transitions controlled only by the external input D-place P_{cp_i}. Further, for each $P_{cp_i} \in P_{cp}$, a predicate transformer $cwlp_{P_{cp_i}}$ on B is defined as follows:

$$cwlp_{P_{cp_i}}(Q)(M) = \begin{cases} 1 & \text{if} \quad T(P_{cp_i}) = \phi \ \text{or} \ wlp_{T_{b_j}}(Q)(M) = 1 \ (\forall T_{b_j} \in 2^{T(P_{cp_i})}) \\ 0 & \text{otherwise} \end{cases}$$

$$(6)$$

The equation (6) implies that $cwlp_{P_{cp_i}}(Q)(M)$ is a transformer that is 1 if several transitions T_{b_j} controlled only by the external input D-place P_{cp_i} at the marking M are concurrently marking-disabled or are transitioned to a state satisfying Q by simultaneous firing of T_{b_j}, and 0 otherwise.

Let $M^{cwlp_{P_{cp_i}}}$ be a set of reachable markings $M \in R(H^{tim})$ such that $cwlp_{P_{cp_i}}(Q)(M) = 1$ for each $P_{cp_i} \in P_{cp}$. Then, a **basis feedback** $b_{P_{cp_i}} : R(H^{tim}) \to \Gamma$ is defined as follows:

$$b_{P_{cp_i}}(M) = \begin{cases} \{P_{cp_i}\} & \text{if} \quad M \in M^{cwlp_{P_{cp_i}}} \\ \phi & \text{otherwise} \end{cases} \quad (7)$$

The equation (7) implies that $b_{P_{cp_i}}(M)$ is a feedback that determines the insertion of only one token into the external input D-place P_{cp_i} for the marking M such that $cwlp_{P_{cp_i}}(Q)(M) = 1$.

[Lemma 9.3] When a predicate Q is control-invariant, for an arbitrary $P_{cp_i} \in P_{cp}$, $b_{P_{cp_i}} \in F(Q)$.

(Proof) Let a basis feedback for $P_{cp_i} \in P_{cp}$ be f. Then, $f = b_{P_{cp_i}}$. It is sufficient to prove that the equation (4) is satisfied for an arbitrary $T_{b_j} \in T_b$. When $T_{b_j} \cap T_c = \phi$, namely, $T_{b_j} \subseteq T_u$, the control-invariance of Q allows that:

$$Q \leq wlp_{T_{b_j}}(Q) \leq wlp_{T_{b_j}}(Q) \vee \neg f_{T_{b_j}} .$$

Let $M \in R(H^{tim})$ be an arbitrary reachable marking. When $T_{b_j} \cap T_c \not\subset T(P_{cp_i})$ or $M \notin M^{cwlp \, P_{cp_i}}$, $\neg f_{T_{b_j}}(M) = 1$ from the equation (7). Therefore, it is sufficient to consider the case of $\phi \neq T_{b_j} \cap T_c \subseteq T(P_{cp_i})$ and $M \in M^{cwlp \, P_{cp_i}}$. It is self-evident from the equation (3) that the equation (4) holds if $D_{T_{b_j}}(M) = 0$. Let us consider the case of $D_{T_{b_j}}(M) = 1$. Let us introduce the decomposition $T_{b_j} = T_{b_jc} \cup T_{b_ju}$. Here, $T_{b_jc} \subseteq T_c$ and $T_{b_ju} \subseteq T_u$. Since $cwlp_{P_{cp_i}}(Q)(M) = 1$ from $M \in M^{cwlp \, P_{cp_i}}$, it is found that $wlp_{T_{b_jc}}(Q)(M) = 1$. Therefore, $Q(M') = 1$ at each reachable marking M' by firing of T_{b_jc} within a small enough time. Further, due to the control-invariance of Q, $wlp_{T_{b_ju}}(Q)(M') = 1$. Hence, the equation (4) is satisfied. Q.E.D.

[Theorem 9.1] When a predicate Q is control-invariant, the following three conditions are equivalent:

(1) The MPF exists.

(2) For an arbitrary $T_{b_j} \in T_b$, arbitrary $f, g \in F(Q)$ satisfy the following:

$$Q \leq wlp_{T_{b_j}}(Q) \vee f_{T_{b_j}} \vee g_{T_{b_j}} \vee \neg(f + g)_{T_{b_j}}$$

(3) For an arbitrary $T_{b_j} \in 2^{T_c} \cap T_b$, the following holds:

$$Q \wedge \bigwedge_{P_{cp_i} \in {}^{\bullet}T_{b_j}} cwlp_{P_{cp_i}}(Q) \leq wlp_{T_{b_j}}(Q)$$
$$(8)$$

118

(Proof) The equivalence of the conditions (1) and (2) is obvious from Lemma 9.2. Hence, let us show that the condition (2) holds when the condition (3) is valid. Suppose that the condition (2) does not hold. Then there exist $f, g \in F(Q)$, $M \in R(H^{tim})$ and $T_{b_j} \in T_b$ such that:

$$Q(M) = 1 \text{ and } (wlp_{T_{b_j}}(Q) \vee f_{T_{b_j}} \vee g_{T_{b_j}} \vee \neg(f+g)_{T_{b_j}})(M) = 0 \ (9)$$

Since Q is control-invariant, it is possible to assume without loss of generality that $T_{b_j} \subseteq T_c$. Let us consider $P_{cp_i} \in^{cp} T_{b_j}$. If $T(P_{cp_i}) = \phi$, $cwlp_{P_{cp_i}}(Q)(M) = 1$ from the equation (6). Let us consider the case of $T(P_{cp_i}) \neq \phi$. Since $(f+g)_{T_{b_j}}(M) = 1$, either $f(M)(P_{cp_i}) = 1$ or $g(M)(P_{cp_i}) = 1$. Since $f, g \in F(Q)$, $wlp_{T'_{b_j}}(Q)(M) = 1$ for an arbitrary $T'_{b_j} \in 2^{T(P_{cp_i})}$. Thus, $cwlp_{P_{cp_i}}(Q)(M) = 1$. Therefore, from the equation (8), $wlp_{T_{b_j}}(Q)(M) = 1$. But this contradicts the equation (9).

Next, let us show that the condition (3) holds if the condition (2) is satisfied. Let us assume that the equation (8) does not hold. Then there exist $M \in R(H^{tim})$ and $T_{b_j} \in 2^{T_c} \cap T_b$ such that:

$$Q(M) = 1 \tag{10}$$
$$cwlp_{P_{cp_i}}(Q)(M) = 1 \ (P_{cp_i} \in^{cp} T_{b_j}) \tag{11}$$
$$wlp_{T_{b_j}}(Q)(M) = 0 \tag{12}$$

From Lemma 9.3, $b_{P_{cp_i}} \in F(Q)$ for each $P_{cp_i} \in^{cp} T_{b_j}$. From the equations (7) and (11), $b_{P_{cp_i}}(M) = \{P_{cp_i}\}$. Thus, if

$$h = \sum_{P_{cp_i} \in^{cp} T_{b_j}} b_{P_{cp_i}}(M)$$

then $h(M) =^{cp} T_{b_j}$, so that $h_{T_{b_j}}(M) = 1$. On the other hand, from Corollary 9.1, $h \in F(Q)$. Therefore, $wlp_{T_{b_j}}(Q)(M) = 1$, which contradicts the equation (12). Q.E.D.

The above discussion holds under the observation of partial states[5].

Example of Not Optimal Design Architecture

Let us consider the existence of the MPF in controlled timed hybrid Petri nets with external input D-places shown as in Fig.9.9. The initial marking is $M_0 = (0,0)$. Let the control specification be $0 \leq M(P_2) \leq 4$. At a marking $M = (4,3)$ reachable from M_0, $Q(M) = 1$. The marking-enabled transitions at M are T_1, T_2 and T_3. Hence, for T_1, $cwlp_{P_{cp1}}(Q)(M) = 1$ ($P_{cp_1} \in^{cp} T_1$) and for T_2, $cwlp_{P_{cp2}}(Q)(M) = 1$ ($P_{cp_2} \in^{cp} T_2$). Since $wlp_{\{T_1, T_2, T_3\}}(Q)(M) = 0$, the equation (8) does not hold. Hence, the MPF does not exist.

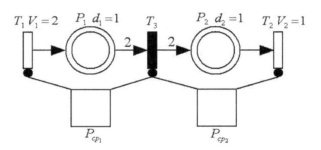

Figure 9.9 Not Optimal Design Architecture

PART TWO

BUSINESS MODELING CONTEXTUALIZATION

Chapter 10. Retail

Selling Business

Recently, UML is actively used for retailing in Japan[1]. Especially in a modern retailer, such as a convenience store, a sales strategy is planned by collecting and exploiting customer's information with a **Point of Sales System**, or for short a **POS system**.

A selling business is described by the sequence diagram shown as in Fig. 10.1. A customer hands a clerk an item at first, then the clerk inputs item data to a checkout counter. After the total price is displayed on the counter, the clerk demands payment of the customer. The customer pays in cash, and the clerk inputs cash data to the counter. At last, the clerk hands the customer the item, change and selling details which are printed out from the counter.

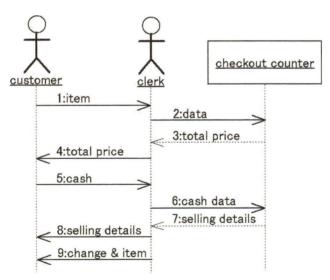

Figure 10.1 Sequence Diagram for Selling Business

A clerk business with a checkout counter is described by the use case diagram shown as in Fig.10.2. These use cases are classified by abstract use cases input and output. As for input, there are use cases to input a bar code or a price, to input a category of a clientele and to input cash data. Also, there are other use cases, as to output, to display a total price and to print out selling details.

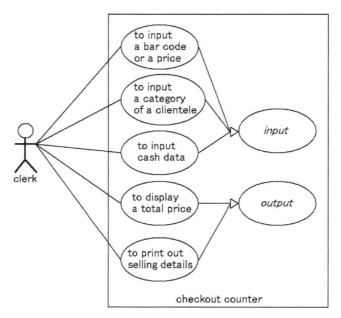

Figure 10.2 Use Case Diagram for Checkout Counter

Supervision of the business data is described by the class diagram shown as in Fig.10.3. Sales records are indicated by the class Sale. The Sale has associations with the classes Store, Employee and Checkout Counter. Also, Sale includes an object of the class Selling Details. The Selling Details has an association with the class Merchandise.

A class has an identification code as an attribute in business modeling, while there is no identifier in real-world modeling which will be discussed later. An object of the Merchandise expresses an item sold and is identified by the **European Article Numbering code**, or for short the **EAN code** which is the world standard bar code. The EAN code has mainly developed in Europe. There are **EAN-13** and **EAN-8** which are a 13-digit and a 8-digit code, respectively.

Also, the **Uniform Product Code**, or for short the **UPC** has mainly developed in the United States of America. Although the UPC is a 12-digit code, the UPC is equivalent to the EAN-13. The **Japan Article Numbering code**, or for short **JAN code** is a subset of the EAN code as well as the UPC.

Note that the multiplications – 1 and 2..* of the association between the Store and the Checkout Counter implies that there is more than one checkout counter at each store, because while a checkout counter is out of order, another counter is used.

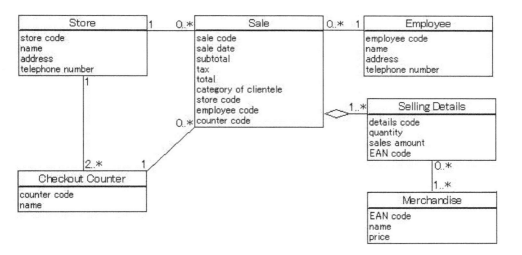

Figure 10.3 Supervision of Business Data

As shown in Fig.10.2, a client inputs category data of a clientele to a checkout counter. Due to the category data, items fit for a clientele can be ready. For example, there are many teenagers as customers, we should arrange items fit for teenagers. Improvement in sales can be achieved by filling a gap between needs of a business district and an arrangement of items at each store. Categories of a clientele are indicated by subclasses of the class Sale such as the Child, the Teenager, the 20th Generation Male, the 20th Generation Female, the 30 to 50th

Generation Male, the 30 to 50th Generation Female, the Advanced-age Male and the Advanced-age Female shown as in Fig.10.4.

Note that client's correct observation and judgment are required for this kind of input. For example, at a convenience store, a part-timer pushes a button which is the easiest to push. This is why all category data of a clientele is advanced in age. Therefore, a system without correct usage could have a bad influence on management even if the system has high functionalities.

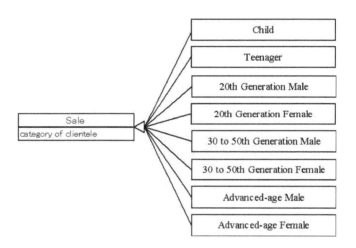

Figure 10.4 Category of Clientele

Sales can be also improved by the device of items offered called **merchandising**. It said in a newspaper that "The cause was the only thing that we were not familiar enough with merchandising." the chief executive officer of a famous retailing company in Japan said this, about the prospect of decrease in income and decrease of profit.

Items should be classified for the sake of merchandizing so that there will be four-layer classes such as the Large Classification, the Middle Classification, the Small Classification and the Merchandise shown as in Fig.10.5. The Large Classification implies a classification for a sales floor such as a grocery, a clothing or a household appliance floor. The Middle Classification implies a classification for a sales department. For example, there is a meat or an alcohol department on a grocery floor. The Small Classification implies a kind of item in a department. For example, there is beer, wine or whisky in an alcohol department. Naturally, the Merchandise implies a distinct item which is identified by an EAN code from the same kind of item. For example, as to beer, there are ANCHOR STEAM BEER, LAGER and ESB.

Sales performance for each object of the class Merchandise should be supervised carefully, because the times have changed from the **mass production** to the **limited production with a wide variety**. We also need to supervise item data with the classes Large Classification, Middle Classification and Small Classification in order to balance the whole merchandise.

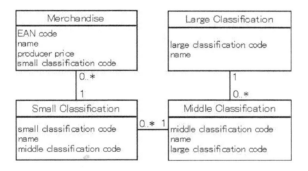

Figure 10.5 Classification for Merchandising

In accordance with the limited production with a wide variety, the class Merchandise should be supervised by various item categories. For example, the Merchandise is categorized by price as the subclasses Main Item, Luxury Item and Bargain Item shown as in Fig.10.6. The Merchandise is categorized by its purpose as the subclass Self-consumption Item or Gift Item shown as in Fig.10.7. The Merchandise is categorized by its season as the subclass Seasonal Item and

Seasonless Item shown as in Fig.10.8. The Merchandise is categorized by its promotion as the subclass Injected New Item, Fast-selling Item or Clearance Sale Item shown as in Fig.10.9.

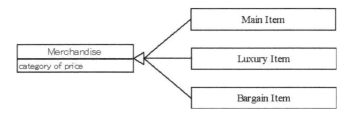

Figure 10.6 Category of Price

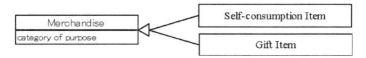

Figure 10.7 Category of Purpose

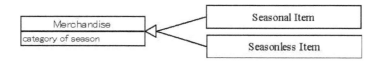

Figure 10.8 Category of Season

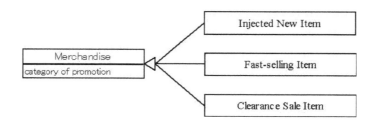

Figure 10.9 Category of Promotion

Order Business

A class with a notation stereo type **<<actor>>** can be used as an actor[2]. Then, intra-office actors clerk, buyer, accountant and manager are expressed as subclasses which inherit from the actor Employee shown as in Fig.10.10. In other words, we can supervise data as to intra-office actors by the class Employee.

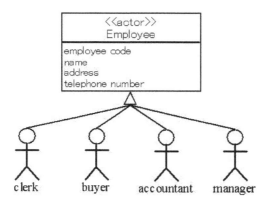

Figure 10.10 Intra-office Actor

In order business, a buyer places an order for an item to a supplier. Suppose a supplier is an actor outside of a company, an order business is described by the use case diagram shown as in Fig.10.11. Order quantity is determined by sales forecast and performance and run out date. Also, order data is sent to a supplier. The mechanism to send and receive data between enterprises is called **electronic data interchange**, for short **EDI**.

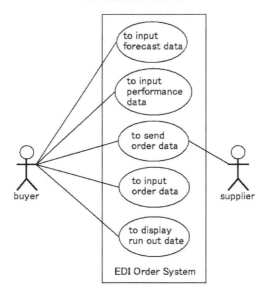

Figure 10.11 Use Case Diagram for Order Business

We consider the purchase index, or for short the PI as follows:

$$PI = \text{total items} \div \text{total customers} \times 100(\%).$$

This equation implies the mean number of items per one customer, independently of the value of items. Note that the loss of sales opportunity by out of stock might decrease the value of PI, thus we should check run out date. Sales forecast is obtained by:

$$\text{sales forecast} = \text{total customers} \times PI \div 100.$$

Thus, order quantity is obtained by:

$$\text{order quantity} = \text{sales forecast} - \text{total stocks}.$$

Supervision of order data is described by the class diagram shown as in Fig.10.12. The class Order has associations with the classes Store, Employee and Supplier, and includes the class Ordering Details.

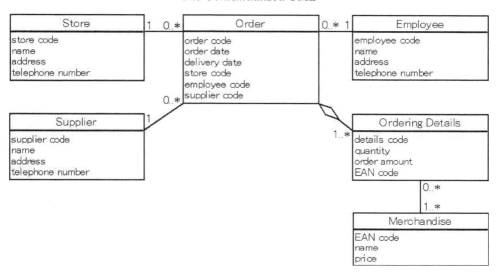

Figure 10.12 Class Diagram for Order Business

Supervision of daily sales forecast and performance is described by the class diagram shown as in Fig.10.13.

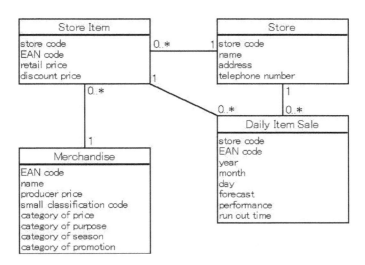

Figure 10.13 Forecast and Performance for Daily Item Sale

Supervision of daily small classification sales forecast and performance is described by the class diagram shown as in Fig.10.14.

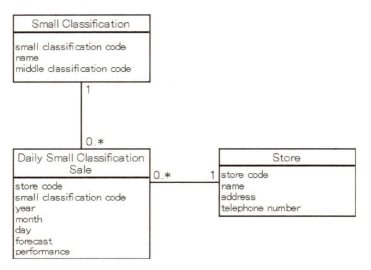

Figure 10.14 Forecast and Performance for Daily Small Classification Sale

Yearly, monthly, weekly and daily total customers forecast and performance are supervised by the class diagram shown as in Fig.10.15. The class Store includes the class Yearly Customer, and the Yearly Customer has the class Monthly Customer. Further, the Monthly Customer includes the class Weekly Customer, and the Weekly Customer has the class Daily Customer.

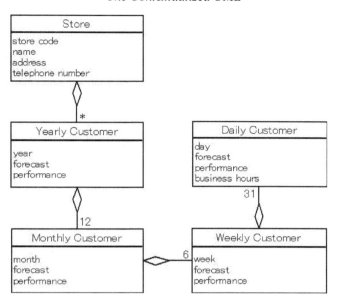

Figure 10.15 Forecast and Performance for Customer

Electronic Commerce Service

Recently, retailing business area has extended by virtue of **electronic commerce**, or for short **EC** on the internet or the world wide web. For example, an evolution of an order system on EC is described by the sequence diagram shown as in Fig.10.16. A customer searches an item on the web, and an e-mail which is written about an order confirmation is sent to the customer in the time of the determination for ordering an item.

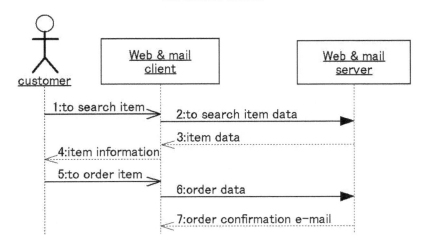

Figure 10.16 Sequence Diagram for EC Order System

A business model, called **Business to Consumer**, or for short **B2C**, is described by the object diagram shown as in Fig.10.17. A customer orders an item on a Web shop. Then, a clerk confirms the order and sends the item to an express company. The express company receives the item from the clerk and the order data from the Web shop. Then, at the express company, the order is confirmed and the item is sent to the customer.

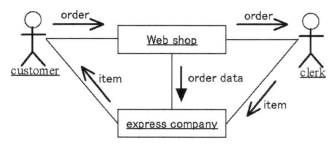

Figure 10.17 Business to Consumer EC Model

Business which processes an order is described by the activity diagram shown as in Fig.10.18. By taking stock according to an order data, an item is shipped if in stock, and it is necessary to adjust stock if out of stock.

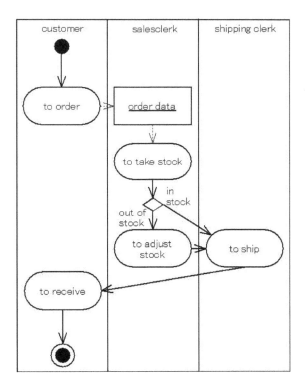

Figure 10.18 Business Process for EC Order

This kind of EC order system is described by the use case diagram shown as in Fig.10.19. There are use cases to search an item, to display an item, to order an item and to display orders. It is, however, necessary to join a member in order to utilize use cases to order. Due to the customer registration, we can keep customer information under perfect control, then sales promotion to a distinct customer is realized. This kind of marketing is said to be **one-to-one marketing**, compared to the conventional **mass marketing**.

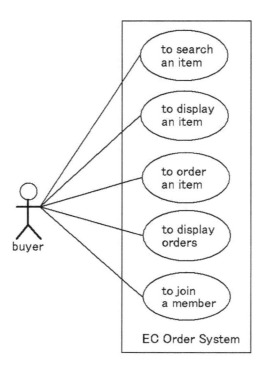

Figure 10.19 Use Case Diagram for EC Order System

At last, Supervision of data in an EC order system is described by the class diagram shown as in Fig.10.20. The class Order has associations with the classes Store and Customer and includes the class Ordering Details. Note that the Order does not have associations with classes such as the Employee and Checkout Counter. Also, the Customer has an e-mail address so that an order can be confirmed by an e-mail.

An order could be canceled before shipping an item. Thus, on the association between the Order and the Sale, the multiplication for the Sale is 0..* where 0 implies the cancel of an order, and * implies more than one sale per one order.

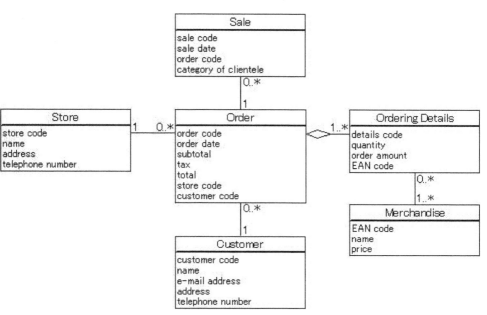

Figure 10.20 Class Diagram for EC Order System

Chapter 11. Wholesale

Wholesaler

Nowadays, UML is used for modeling wholesale in Japan[1]. The role of wholesale in physical distribution is described by the communication diagram shown as in Fig.11.1. In the physical distribution, a retailer requires multi-frequent and small sum, and a manufacturer requires few frequent and large sum. Then, a wholesaler fills the role of an intermediary which effectively adjusts the physical distribution for different requirements between a retailer and a manufacturer.

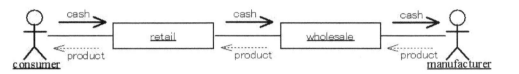

Figure 11.1 Communication Diagram for Wholesale in Physical Distribution

A wholesale which satisfies requirements of a retailer and a manufacturer is described by the use case diagram shown as in Fig.11.2. The wholesale has business use cases such as selling business, stocking business, physical distribution and retail support. The physical distribution is shared by the retailer and the manufacturer. The selling business is a business only for the retailer, and the stocking business is a business only for the manufacturer. The retail support is a business in order to offer the retailer useful information.

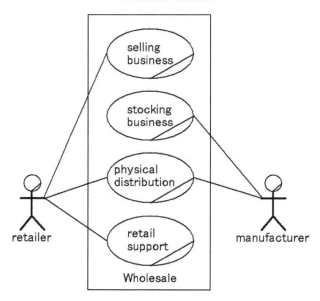

Figure 11.2 Use Case Diagram for Wholesale

Selling Business

Selling business has use cases to receive an order, to receive an inquiry and to collect a bill which are described by the use case diagram shown as in Fig.11.3. We have three general ideas how an order is received: to visit, to call or fax and especially the efficiency of wholesale business which has been increased by introducing **electronic ordering system**, or for short **EOS**.

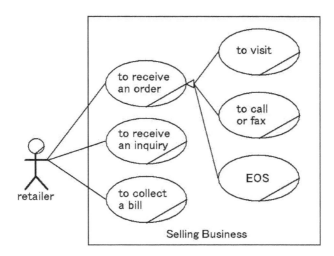

Figure 11.3 Use Case Diagram for Selling Business

Receiving an order by EOS is described by the activity diagram shown as in Fig.11.4. At first, a retailer inputs order data into an order client computer, and sends the order data to an order server computer of wholesale. Next, the order server checks the order data. Then, the correct data is registered and the error data is sent to the order client of retail. The received error data is printed out by the order client computer.

—

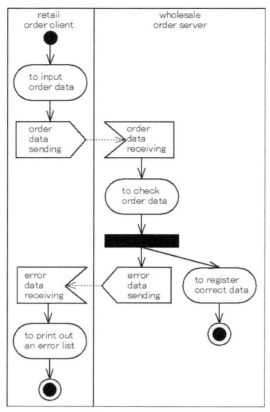

Figure 11.4 Activity Diagram for EOS

Vendor Managed Inventory

As for EOS, there is **vendor managed inventory**, or for short **VMI** which is described by the sequence diagram shown as in Fig.11.5. A retailer sends sale and inventory data to a wholesaler. The wholesaler judges the necessity for shipment only from the data without order data. In shipping, the wholesaler sends **advanced shipping notice data**, or for short **ASN data**, and then ships items to the retailer. Note that the wholesaler has the ownership for the shipped items. And the sales of the wholesaler are added up simultaneously with the sales of the retailer. At last, the bill data are sent from the wholesaler to the retailer on a day decided beforehand.

Such a computerized reliable partnership is called **supply chain management**, or for short **SCM**.

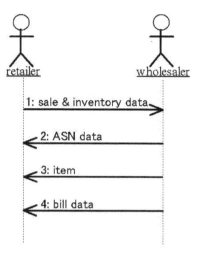

Figure 11.5 Sequence Diagram for VMI

In selling business by VMI, the use case to receive sale and inventory data, which is a function peculiar to VMI, is added to the use case diagram, shown as in Fig.11.6. Namely, the retailer can receive sale and inventory data from the wholesaler. Also, there are use cases as to sales appropriation and inventory supplement which are inherited from the function to receive sale and inventory data.

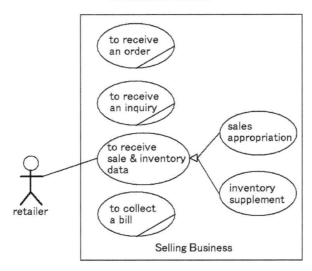

Figure 11.6 Use Case Diagram for Selling Business by VMI

The use case to receive sale and inventory data is described by the activity diagram shown as in Fig. 11.7. At first, a retailer sends sale and inventory data to a wholesaler. When the wholesaler receives the data, sales appropriation is conducted parallel with inventory supplement only if out of stock.

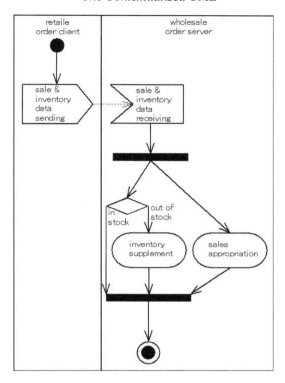

Figure 11.7 Activity Diagram for Receiving Sale & Inventory Data

Stocking Business

Relationship between a manufacturer and stocking business in wholesale is described by the use case diagram shown as in Fig.11.8. The stocking business has ordering business, acceptance inspection and payment. In the ordering business, the wholesaler receives an item from the manufacturer. In the acceptance inspection, the received item is confirmed and registered. In the payment, the price for the received item is paid.

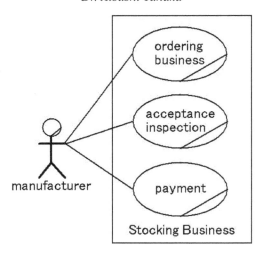

Figure 11.8 Use Case Diagram for Stocking Business

In the ordering business, there are two types of order system: a **fixed-quantity order system** and a **fixed-period order system**. The former is a system that fixes the amount of orders and adjusts the timing of orders. Then, an order will be placed early if the number of stock decreases early, and the timing of an order is late if the number of stock decreases slowly. The latter is a system that fixes the timing of orders and adjusts the amount of orders. Then, an order will be placed mostly if the number of stock decreases early, and the amount of orders is decreased if the number of stock decreases slowly.

Thus, the amount or the timing of orders depends on distributor's stock at a distribution center. Supervision of data by such order systems is described by the class diagram shown as in Fig.11.9. The class Merchandise has association with the class Distributor's Stock, and the Distributor's Stock has association with the class Distribution Center. Further, the class Order has association with the classes Employee and Manufacturer, and includes the class Ordering Details. And the Ordering Details has association with the Merchandise.

```
┌──────────────┐1    0..*┌──────────────┐0..*    1┌──────────────┐
│   Employee   │─────────│    Order     │─────────│ Manufacturer │
└──────────────┘         └──────────────┘         └──────────────┘
                                │
                                ◇
                               1..*
                         ┌──────────────┐0..*  1..*┌──────────────┐
                         │Ordering Details│───────│ Merchandise  │
                         └──────────────┘         └──────────────┘
                                                         │1
                                                         │
                                                         │0..*
                    ┌──────────────┐1    0..*┌──────────────┐
                    │ Distribution │─────────│ Distributor's│
                    │   Center     │         │    Stock     │
                    └──────────────┘         └──────────────┘
```

Figure 11.9 Class Diagram for Order System

Acceptance Inspection

Acceptance inspection in wholesale is business at a distribution center, which is described by the use case diagram shown as in Fig.11.10. In the acceptance inspection, the contents and quantity of a load are confirmed by scanning a bar code called the **interleaved two of five code**, or for short **ITF code** which is an extended EAN code.

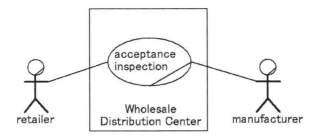

Figure 11.10 Acceptance Inspection at Distribution Center

Automated acceptance business is based on a computerized reliable partnership, between a manufacturer and a wholesaler, called an **inspectionless system**, which is described by the activity diagram shown as in Fig.11.11. In this system, an ASN data indicating the contents and quantity of a load on a pallet which is a load-carrying platform, is sent from a manufacturer to a wholesaler. And after that, the load is shipped. After the wholesaler receives the load, a **supply chain management label**, or for short a **SCM label** is attached to a pallet on which the received load is put. The pallet is identified by the SCM label. Thus, the SCM label is only scanned without acceptance inspection, because the received ASN data is reliable in inspectionless systems. Then, a seal for inventory management is affixed to the load and the load is put away in a depository at last.

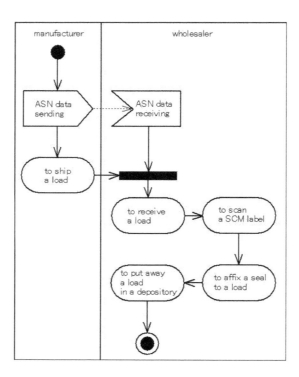

Figure 11.11 Activity Diagram for Inspectionless System

Supervision of data by acceptance inspection is described by the class diagram shown as in Fig.11.12. Acceptance is recorded by the class Arrival Notice associated with the class Distribution Center. The Arrival Notice has association with the classes Order and ASN. And the multiplication of Arrival Notice – 0..* implies that association with the Order or the ASN is acceptance inspection or inspectionless system, respectively. Also, storage of items is indicated by the association among the classes Merchandise, Depository, Packing and Arrival Notice.

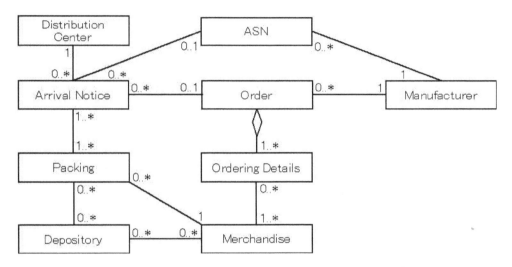

Figure 11.12 Class Diagram for Acceptance Inspection

Comprehensive Physical Distribution System

In traditional ways of thinking, physical distribution has the flow of items from a manufacturer to a retail store. Recently, however, a **comprehensive distribution system**, which delivers for every items according to shelf to a retail store, has been in use. This system is described by the communication diagram shown as in Fig.11.13.

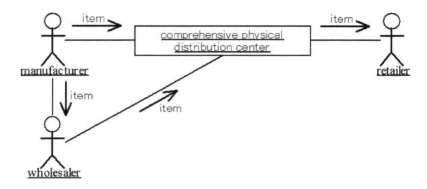

Figure 11.13 Communication Diagram for Comprehensive Physical Distribution

In a comprehensive distribution center, picking business which takes out items from a warehouse according to orders from a retailer, is significant for wholesale. There are **single picking** which takes out items for each distinct store, and **total picking** which collectively takes out items for all stores.

Supervision of data by picking business is described by the class diagram shown as in Fig.11.14. There is an association between the classes Picking and Order, because picking business is conducted by receiving orders from a retailer. Then, the multiplication of the Order – 1..* implies single picking if the multiplication is one, and total picking if the multiplication is more than one. Further, the Picking also has association with the class Distribution Center, which implies which distribution center the location of picking is.

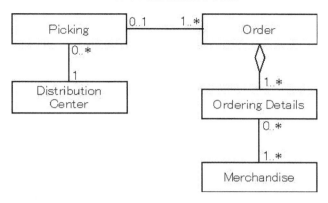

Figure 11.14 Class Diagram for Picking Business

Sending and receiving data for supervision of physical distribution is described by the sequence diagram shown as in Fig.11.15. An order data is sent from a retailer to a chain store headquarters. The headquarters puts together order data from retailers and sends the data to wholesalers. Then, delivery data is sent from the headquarters to a comprehensive physical distribution center. This data is used at the distribution center for picking business according to shelf of stores. Also, an ASN data is sent from the wholesaler to the distribution center. And at the distribution center, acceptance without inspection is done by the ASN data. This ASN data is also sent from the distribution center to a retail store to do acceptance without inspection at the store. At last, bill data is sent from the wholesaler to the headquarters, and paying notice data is replied from the headquarters to the wholesaler.

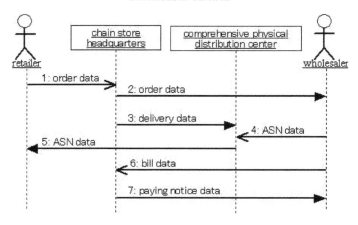

Figure 11.15 Sequence Diagram for Comprehensive Physical Distribution System

Chapter 12. Discussion

Real-world Modeling

Modeling classes in real-world is said to be **real-world modeling**[1].

[Definition 12.1] Real-world modeling is defined by:

(i) A class corresponding with a set of objects in real-world is identified.

(ii) Association between different classes corresponds to a static relationship. Different sets of objects are identified.

(iii) An attribute of a set of objects is given to a class corresponding to the set, as an attribute of the class.

(vi) A function of a set of objects is given to a class corresponding to the set, as an operation of the class.

Real-world modeling is used in order to analyze real-world in which existing problems are expressed clearly and the solutions are to be considered. For example, real-world modeling is applied to **business process reengineering**, or for short **BPR**.

Also, real-world modeling is used for a computer to simulate real-world such as flight simulator. In this usage, real-world is modeled as cyber-world in a computer. A software tool such as a text editor, a graphic editor or a compiler similarly exists in cyber-world. Thus, real-world modeling is also used in order to develop such software.

Pseudo Real-world Modeling

Real-world modeling cannot be applied to developing a man-made system which automates business process in real-world. In this context, the relation of a man-made system to real-world is shown as in Fig.12.1. Real-world is indicated by Fig.(a). As shown in Fig.(c), a man-made system which implies automated business process is driven by a user. A man-made system consists of three types of class which are identified by the stereotype <<Presentation>>, <<Plant>> and <<Supervisor>>. Thus, the system of Fig.(c) is automated real-world. However, it cannot be obtained by real-world modeling. The system is designed based on classes of Fig.(b). Classes of Fig.(b) correspond to plant classes of Fig.(c). It is said to be **pseudo real-world modeling** to extract classes of Fig.(b) from real-world of Fig.(a)[1].

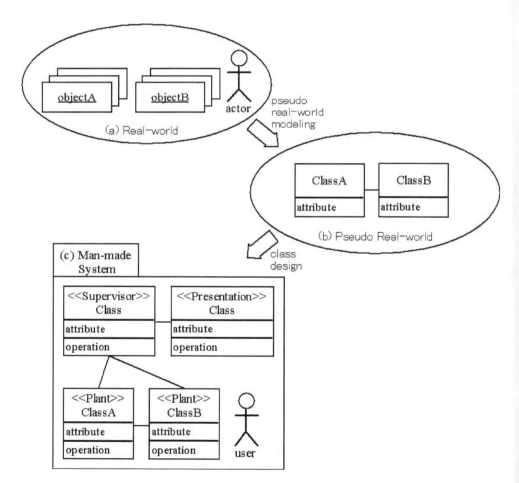

Figure 12.1 Pseudo Real-world Modeling for Designing Man-made System

Pseudo real-world classes of Fig.(b) have an attribute of information processed by a man-made system of Fig.(c), because they correspond with plant classes of Fig.(c). On the other hand, classes of Fig.(b) have no operations in pseudo real-world modeling, because an operation of plant classes depends on how to design supervisor classes in Fig.(c).

[Definition 12.2] Pseudo Real-world modeling is defined by:

(i) A class corresponding with information processed by a man-made system is identified.

(ii) Association between different classes corresponding with different processed information is identified.

(iii) An attribute of processed information is given to a class corresponding to the processed information as an attribute of the class.

For example, let us consider a simple library system where three requirements are as follows[1]:

(i) A user checks out or returns a book.

(ii) A librarian registers a book.

(iii) Data on a book and a user who borrows the book are registered.

From the requirement (iii), there are a book class, a user class and their association on which a user borrows a book, as shown in Fig.12.2. A user class has attributes such as user code and user name. A book class has attributes such as international standard book number, or for short ISBN, book name and a rental code which shows whether the book is borrowed or not.

Dr. Atsushi Tanaka

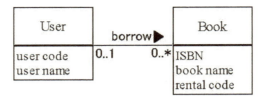

Figure 12.2 Pseudo Real-world Modeling for Simple Library System

In a design process, the class diagram of Fig.12.3. is obtained from the classes of Fig.12.2. The class User or Book is a plant class. In the Book, operations such as "borrow", "return" and "register" are added. The "register" accesses the ISBN and the book name. The "borrow" and the "return" access the rental code. Also, in the User, operations such as "check out" and "return" are added. The supervisor class has control specifications of these operations. The requirements (i) and (ii) are expressed by the presentation classes User Window and Librarian Window, respectively.

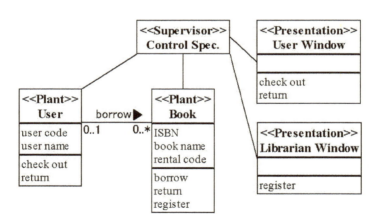

Figure 12.3 Designing Class Diagram for Simple Library System

Now, what happens by applying real-world modeling to the simple library system? By requirements (i), (ii) and (iii), a real-world model is obtained, as shown in Fig.12.4. This model is valid for library simulation. However, the model is erroneous for developing a man-made system to support library business. Especially, the class Librarian is not necessary. And every operation is given regardless of control specifications. Also, unnecessary association corresponding to redundant operation is given.

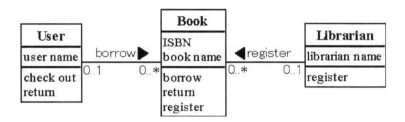

Figure 12.4 Erroneous Modeling for Library System

A communication diagram is shown as in Fig.12.5, in case where actors such as a user and a librarian respectively access classes such as a user and a librarian. This model is not valid, because objects such as the user and the librarian in the system respectively, repeat operations by actors such as the user and the librarian in the real-world. Therefore, real-world modeling is considered harmful for developing the library system.

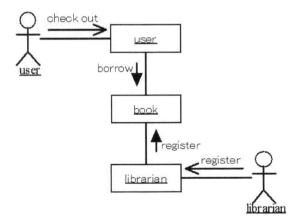

Figure 12.5 Communication Diagram for Erroneously Designed Library System

157

Dr. Atsushi Tanaka

How to Extract Class for Pseudo Real-world Modeling

It is generally difficult to extract classes in pseudo real-world modeling. In conventional methods, we should collect information from business documents at first, because in business processes, flows like an activity diagram are not necessarily well-documented in details. However, it is not difficult to describe their relevant activities by business use cases.

We will now show how to extract class diagrams from business use cases through state machine dynamics from the view point of actors. Due to this method, we can find all required information from the dynamics, even if a part of the necessary information is lacking in information collected from business documents at first. Also, a man-made system developed by the required information is desirable, because such a system has mechanism to realize an appropriate business model for an actor.

As shown in Fig.12.6, we consider an example of a use case diagram for a school information system[2]. A student attends a lecture after reporting the subject to a registrar. After an examination, a teacher reports the scores to a registrar. Subjects and scores are registered by a registrar. After the registrations, they can be seen by the relevant students and teachers.

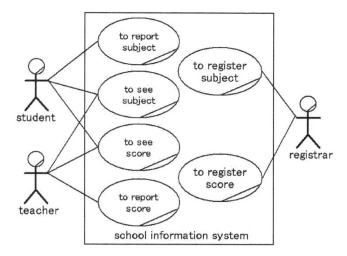

Figure 12.6 Use Case Diagram for School Information System

A state machine shown as in Fig.12.7 is derived from use cases to report a subject and to see a subject and a score by an actor student.

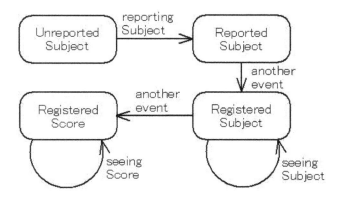

Figure 12.7 State Machine by Student

A state machine shown as in Fig.12.8 is derived from use cases to report a score and to see a subject and a score by an actor teacher.

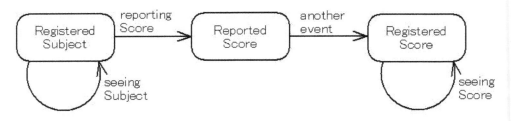

Figure 12.8 State Machine by Teacher

A state machine shown as in Fig.12.9 is derived from use cases to register a subject and a score by an actor registrar.

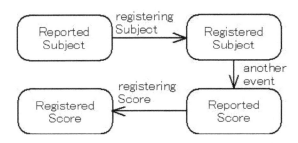

Figure 12.9 State Machine by Registrar

As shown in Fig.12.10, two classes, Student and Subject and their association are extracted from the state machine reporting a subject.

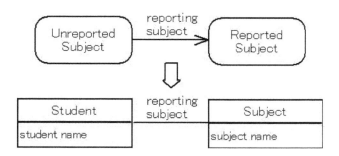

Figure 12.10 Class for Subject by Student

As shown in Fig.12.11, two classes, Registration and Subject and their association are extracted from the state machine registering a subject.

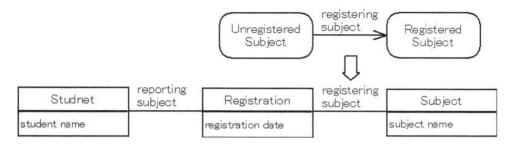

Figure 12.11 Class for Subject by Registrar

As shown in Fig.12.12, a class Score is extracted from the state machine reporting a score.

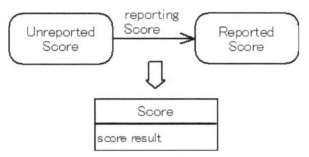

Figure 12.12 Class for Score by Teacher

As shown in Fig.12.13, two classes, Registration and Score and their association are extracted from the state machine registering a score.

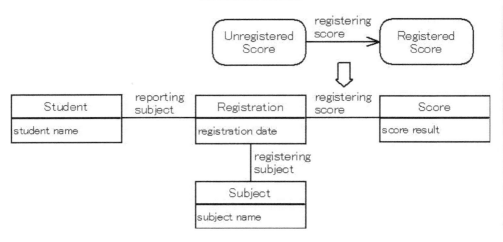

Figure 12.13 Class for Score by Registrar

Finally, the class Registration is changed to an association class, then a class diagram is derived as shown in Fig.12.14.

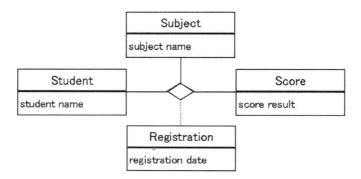

Figure 12.14 Extracted Class Diagram

Component Analysis

A **component-based development** is an extremely efficient development, because this method shows how to assemble software parts[3]. Thus, it is important to reuse an existing component. As a matter of fact, components are, however, not reused so much as software parts.

So that a component is reused, we extract a component from a business process and define its specification which is especially interface of the component in the early stage of development processes. Then we select an appropriate component from exchangeable candidates of components adaptive to the specification.

For example, we consider developing an UML model management system. First, we deal with a business process using the system described by the activity diagram, as shown in Fig.12.15 where modeling, registering, searching, updating or deleting are repeated.

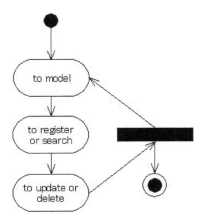

Figure 12.15 Activity Diagram for UML Model Management System

From the activities, a management component has interfaces to register, search, update and delete. Further, we define another component with an interface to display. Then, the management component implements the interface to display. Therefore, the components Display and Management are derived as shown in Fig.12.16.

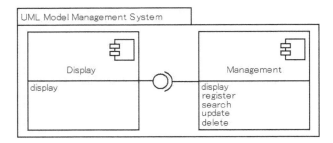

Figure 12.16 Extracted Component

Business Transaction Pattern

In business modeling contextualization, business processes suitable for component design could be expressed by certain patterns. And it can be easy to analyze and design components through the patterns. Such a pattern is said to be a **business transaction pattern**[4]. The typical patterns are presented as follows[4].

A **query/response pattern** is applied to a business process where a requester obtains information which a respondent has. For example, a business process where a buyer obtains catalogue which a seller has is described by the activity diagram as shown in Fig.12.17.

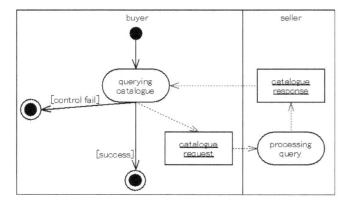

Figure 12.17 Query/Response Pattern for Querying Catalogue

The Contextualized UML

A **request/response pattern** is applied to a business process where a requester obtains information which is peculiar to the requester from a respondent. For example, a business process where a buyer obtains number which identifies the buyer from a seller is described by the activity diagram as shown in Fig.12.18.

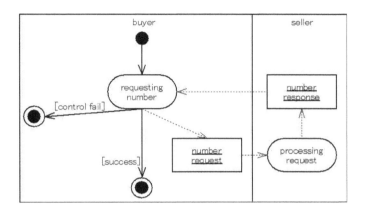

Figure 12.18 Request/Response Pattern for Requesting Number

A **request/confirm pattern** is applied to a business process where a requester confirms information which a respondent has. For example, a business process where a buyer confirms an order to a seller is described by the activity diagram as shown in Fig.12.19.

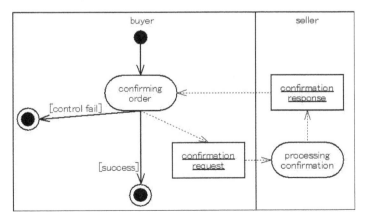

Figure 12.19 Request/Confirm Pattern for Confirming Order

165

An **information distribution pattern** is applied to a business process where a requester sends information to a respondent. For example, a business process where a buyer sends an item reservation to a seller is described by the activity diagram as shown in Fig.12.20.

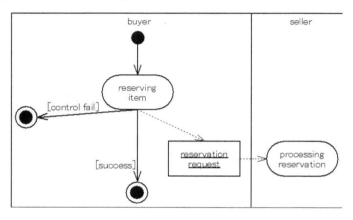

Figure 12.20 Information Distribution Pattern for Reserving Item

A **commercial transaction pattern** is applied to a business process where a contract document of a requester is accepted by a respondent in business transaction. For example, a business process where a sales contract is signed between a buyer and a seller is described by the activity diagram as shown in Fig.12.21.

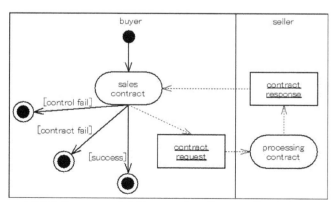

Figure 12.21 Commercial Transaction Pattern for Sales Contract

PART THREE

INDUSTRIAL SYSTEM CONTEXTUALIZATION

Chapter 13. Analysis Process

Complicated Industrial System

A computer system which is embedded in an equipment for the purpose of control is said to be an **embedded system**. Japan is famous for electronics products. Such products are supported by embedded systems.

In a previous embedded system just as a control system, a software program decides control action according to a state. This program is called **state-driven**. Due to the growth of semiconductor and microprocessor technologies, an embedded system has recently become advanced functional, large scaled and complicated. In such a modern embedded system, a software program decides control action according to a message which is received from a network terminal. This program is called **event-driven**.

For example, a cellular phone or a car navigation system which has function as **wireless networking** or **multimedia processing**, processes sound or images data as well as text or binary data. Also, **graphical user interface**, for short, **GUI**, which is easy to use is required for this kind of system.

For developing a mobile computing system, we have problems such as battery power consumption or the radiation of heat. Then, it is not easy to decide which **hardware/software partitioning** or what portion is realized as either hardware or software. Previously, only one mature engineer designed both hardware and software. This development style is desirable for optimal design. However, large numbers of engineers should share work for large development to be furthered efficiently. As a matter of fact, the development term tends to become shorter regardless of the scale of large development. After all, labor environment has been getting worse in such a dilemma.

This is why we have problems such as increasing of development processes or decreasing of quality. A bottom up approach is based on source program codes, only for the purpose of executing a computer, not effective for this kind of problem. In order to solve the problems, a model-driven approach based on the UML is useful, because it is easier to manage development process using the UML as a communication tool. Especially, we should unite the UML approach in software engineering with analysis and design techniques in system engineering. However, it is not easy to apply the UML for embedded systems. This and the next chapters describe analysis and design of software in an embedded system development. An approach to difficulties on behaviors of complicated industrial systems will be discussed in the last chapter.

Development Domain

To develop complicated systems, the development domain should be divided into different domains in which a special knowledge is required. In general, the development domain of embedded systems is divided into application software domain, system software domain, user interface domain and hardware device domain as shown in Fig.13.1[1]. The dependency of the package diagram implies that application software is executed using user interface and system software which are implemented on hardware devices.

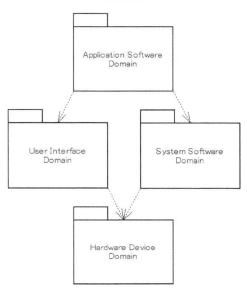

Figure 13.1 Divided Development Domain for Specialist

Application Software Domain

As an example of a simple embedded system, we consider factory automation, for short FA in which an equipment sorts parts of a lot and distinguishes a lot for sorted parts as shown in Fig.13.2.

Figure 13.2 FA Equipment for Sorting Part and Distinguishing Lot

At first, user requirements for application software of FA equipment are described by the business use cases shown as in Fig.13.3. It is required to sort parts according to size and distinguish a lot for parts from user point of view.

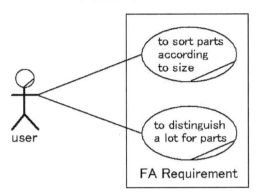

Figure 13.3 Business Use Case for Requirement of FA Equipment

A state machine is derived from the business use cases of Fig.13.3. As shown in Fig. 13.4, sorting parts is described by the state machine. The initial state is idle. The other state is a state which sorts a part. A state for waiting a part is also idle.

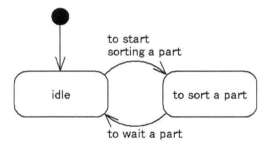

Figure 13.4 State Machine for Sorting Part

As shown in Fig.13.5, distinguishing a lot is described by the state machine. It is stated to distinguish a lot when sorting all parts is completed. After distinguishing a lot, it is started to sort out a part again.

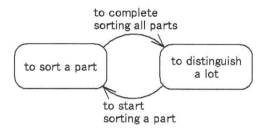

Figure 13.5 State Machine for Distinguishing Lot

The state machine shown in Fig.13.6 is obtained by combining the state machine of Fig.13.4 with the state machine of Fig.13.5.

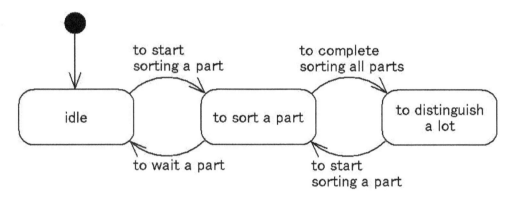

Figure 13.6 State Machine for Business Use Case

The FA equipment is started by a power supply switched on by a pushed button, and is ended by a power supply shut off. Thus, as shown in Fig.13.7, the dynamics of the equipment are hierarchically described by the state machine.

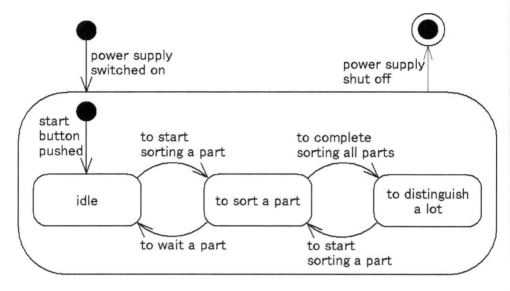

Figure 13.7 State Machine for FA Equipment

System Software Domain

A hardware device of an equipment is controlled by system software. Hardware requirements for system software of the FA equipment are described by the system use cases as shown in Fig.13.8. System software to distinguish a part or convey a part is required for a sensor or a belt conveyor, respectively.

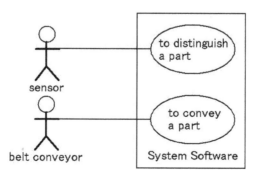

Figure 13.8 System Use Case for System Software of FA Equipment

User Interface Domain

Software peculiar to user interface is required for an interface device such as a button or a display as shown in Fig.13.9. The button and the display collaborate with system use cases to receive input and display information, respectively.

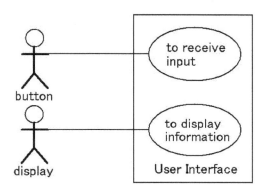

Figure 13.9 System Use Case for User Interface of FA Equipment

Hardware Device Domain

As shown in Fig.13.10, hardware devices are also described by the system use cases. This diagram implies that there are input and output ports as physical interface and their roles are functions to input and output signal, respectively.

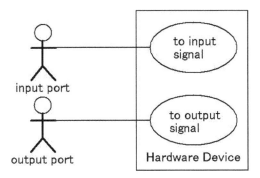

Figure 13.10 System Use Case for Hardware Device of FA Equipment

Communication over Domain

System requirements are obtained from the state machine of Fig.13.7 and the use case diagrams of Fig.13.3, 13.8 and 13.9. They are expressed by communication over such domains: application software, system software and user interface. Then, we describe the communication over package diagram shown as in Fig.13.11. There are only actors in the user interface and the system software domains. The application software domain has sorted result data which are updated and used for distinguishing a lot.

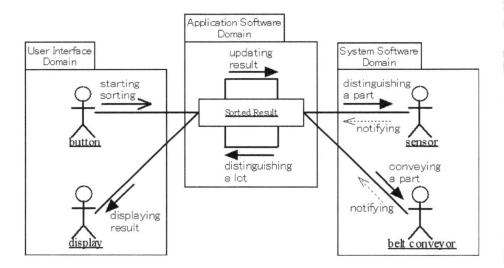

Figure 13.11 Communication over Package Diagram for FA Equipment

Pseudo Real-world Modeling

It is found from Fig.13.11 that the class Sorted Result is required. As shown in the class diagram of Fig.13.12, the Sorted Result has lot number as an attribute. Also, the class Sorted Part is included by the Sorted Result. The Sorted Part has part name and part sum as attributes.

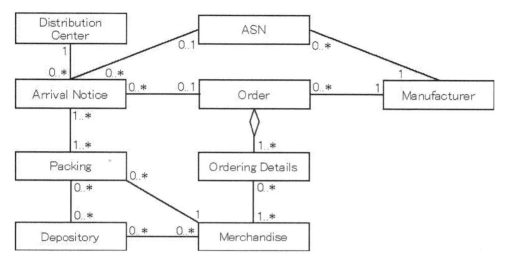

Figure 13.12 Class for Result Data

As shown in the class diagram of Fig.13.13, an object of the class Lot is distinguished by the Sorted Result. The Lot has lot name and part sum as attributes.

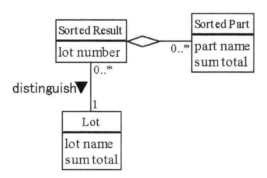

Figure 13.13 Class for Distinguishing Lot

As shown in the class diagram of Fig.13.14, the class Part is included by the Lot. The Part has part name and part sum as attributes.

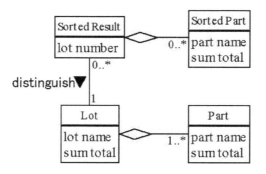

Figure 13.14 Class for Detailed Lot

An object of the Part is distinguished by the Sorted Result. Thus, the class diagram shown as in Fig.13.15 is obtained.

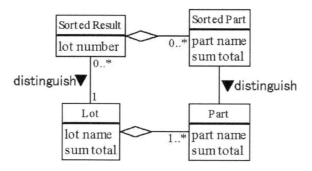

Figure 13.15 Association for Distinguishing Part

At last, each part is related to a conveyance place indicated by the class Conveyance Place in the class diagram of Fig.13.16. The Conveyance Place has location as an attribute. It is expressed by the association between the Part and the Conveyance Place, which is a part conveyed to a location.

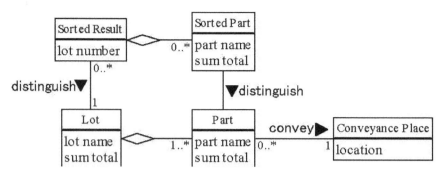

Figure 13.16 Pseudo Real-world Modeling for Developing FA Equipment

Chapter 14. Design Process

Domain Architecture

We consider dependency over development domains in advance of system design[1]. As shown in Fig.14.1, the application software domain depends on the user interface domain through the interface User Interface whose operation is "display", and on the system software domain through the interface System Software whose operations are "distinguish" and "convey".

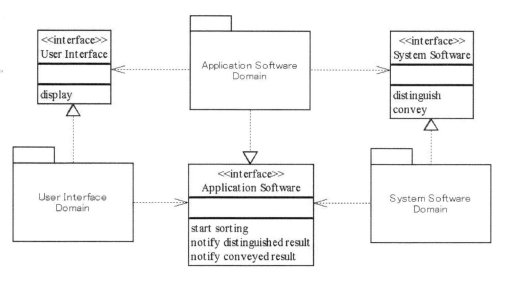

Figure 14.1 Package Diagram from Application Software View Point

This diagram expresses software architecture from application software view point. Then, the interface Application Software is considered harmful for maintenance, because the interface is shared by user interface domain and system software domain. Namely, The application software domain has mixed collaboration with a user and a system. Thus, the direction of dependency from the user interface and the system software domains to the application software domain is changed to the modified direction from the user interface domain to the system software domain through the application software domain. Then, the modified package diagram is obtained as shown in Fig.14.2.

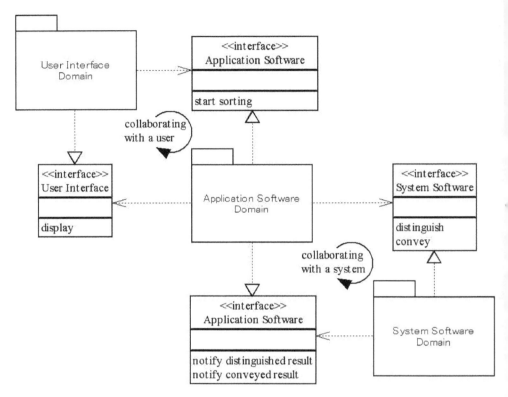

Figure 14.2 Modified Package Diagram for Maintenance

Application Software Design

We consider application software design based on the pseudo real-world modeling of Fig.13.16. At first, the behavior of a supervisor class for sorting is described by the state machine shown as in Fig.14.3. This diagram implies that parts are distinguished and conveyed.

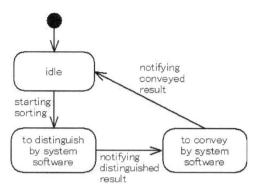

Figure 14.3 State Machine for Supervision of Sorting

According to the state machine diagram of Fig.14.3, the supervisor class Sorting implements two interfaces Application Software and has dependency with the interface System Software as shown in Fig.14.4.

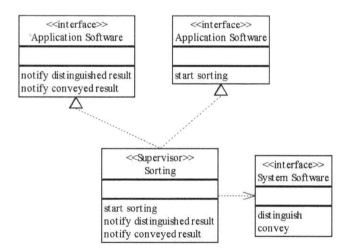

Figure 14.4 Supervisor Class Sorting

Also, the supervisor class Distinguishing is introduced for distinguishing a lot as well as the Sorting. As shown in Fig.14.5, the communication diagram is useful to determine operations of classes. At first, the supervisor class Sorting starts sorting a part. Then, the plant class Sorted Part updates sorted result. And the plant class Part gets conveyance place and the location is gotten by the plant class Conveyance Place. When sorting is completed, sorted result is confirmed by the plant class Sorted Result under control of the supervisor class Distinguishing. Then, the sum of parts is gotten by the plant class Lot, and sorted parts are checked by the Sorted Part. Sorted part sum and name are gotten by the Part, and lot name is gotten by the Lot. At last, the result is displayed in the user interface domain.

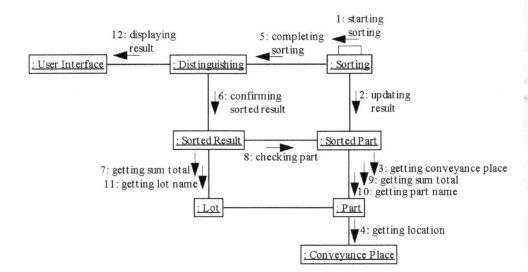

Figure 14.5 Communication Diagram for Application Software Design

An operation in a class diagram corresponds to a message in a communication diagram, which is said to be **responsibility** of a class. Namely, class design can be said to be a process which imposes responsibility on a class. Thus, as shown in Fig.14.6, the class diagram is derived from the communication diagram of Fig.14.5.

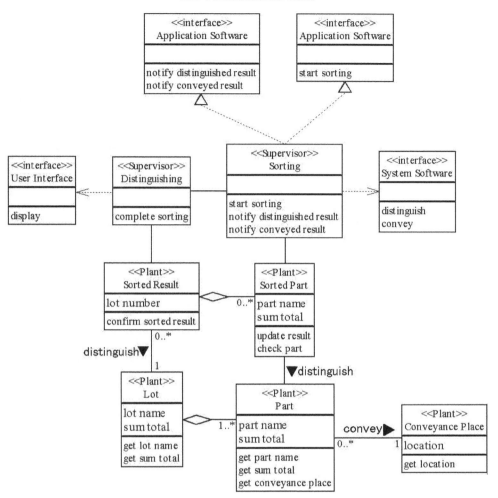

Figure 14.6 Class Diagram for Application Software Design

System Software Design

We consider development in the system software domain. This development process is furthered independently by other development domains. Namely, each domain is developed concurrently. This is generally said to be **concurrent engineering**.

At first, the supervisor class Controller implements the interface System Software and has dependency with the interface Application Software as shown in Fig.14.7. The Controller has operations – distinguish and convey. The distinguish is delegated to the plant class Distinguishing, and the convey is delegated to the plant classes Conveying and Rotating. The Rotating has responsibility to change direction of belt conveyor according to the location of a conveyance place.

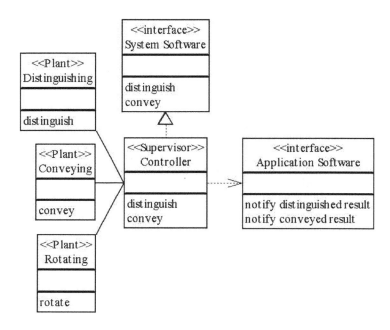

Figure 14.7 Supervisor Class Controller for System Software Design

As shown in Fig.14.8, the plant classes Distinguishing, Conveying and Rotating have association with the other plant classes Sensor, Conveyor Motor and Rotary Motor, respectively. Such a class that has port number as an attribute is software which accesses the interface Hardware Device.

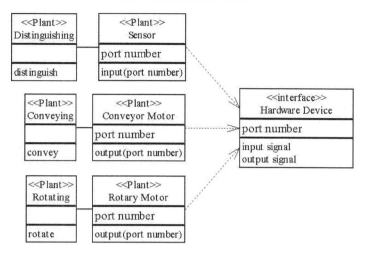

Figure 14.8 Plant Class for Accessing Hardware Device

A part is distinguished according to the size. As shown in Fig.14.9, the class diagram describes or distinguishes a part by checking whether a part image is matched with a graphic pattern, or not. The plant class Graphic Pattern has a graphic pattern image, and the plant class Compensation has the values of upper and lower bounds for the part image as compensation values.

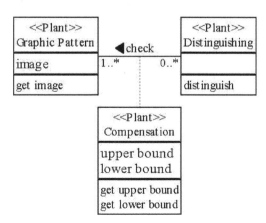

Figure 14.9 Class Diagram for Distinguishing Part

As shown in Fig.14.10, the plant class Conveying has association with the plant class Pushing Out. The Pushing Out has speed as an attribute. This class diagram implies speed adjustment of a belt conveyor.

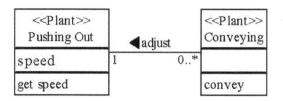

Figure 14.10 Class Diagram for Speed Adjustment of Belt Conveyor

The direction of a belt conveyor which is changed by rotating the conveyor is described by the class diagram shown as in Fig.14.11. The plant class Direction has location as an attribute and reflexive association which implies the left side and the right side locations. Also, the association plant class Rotation has the speed of rotation.

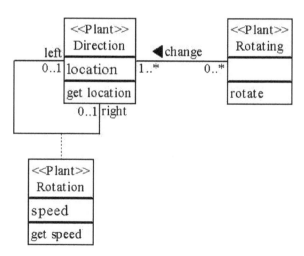

Figure 14.11 Class Diagram for Changing Direction of Belt Conveyor

A procedure of changing direction of the belt conveyor is described by the activity diagram shown as in Fig.14.12. At first, the direction to a conveyance place is decided according to the size of a part. The decided direction is fixed if the direction is the same as current direction of the conveyor, and the direction of the conveyor is changed to the decided direction otherwise.

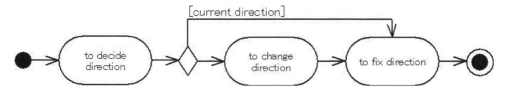

Figure 14.12 Activity Diagram for Changing Direction of Belt Conveyor

A procedure of conveyance is described by the activity diagram shown as in Fig.14.13. A part is pushed out at first and then conveyed. And there is a process for completing the conveyance in the end.

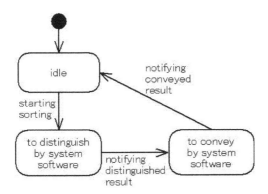

Figure 14.13 Activity Diagram for Conveying Part

Dr. Atsushi Tanaka

In order to convey a part as soon as possible, these procedures are described by the parallel activity diagram shown as in Fig.14.14. Namely, a part is conveyed while changing the direction to the conveyance place.

In general, a trouble may happen during conveyance of parts. A procedure of the recovery is described by the activity diagram shown as in Fig.14.15. When an error occurs during conveyance of a part, the conveyance can be retried as far as the retry count is less than the value of an upper bound, even if the retry fails. When the automatic recoveries do not succeed at last, the conveyance is recovered at hand.

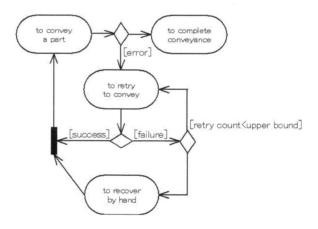

Figure 14.15 Activity Diagram for Error Recovery

The class diagram for error recovery of conveyance, where the plant class Error is added, is shown as in Fig.14.16. The Error has retry count and upper bound as attributes.

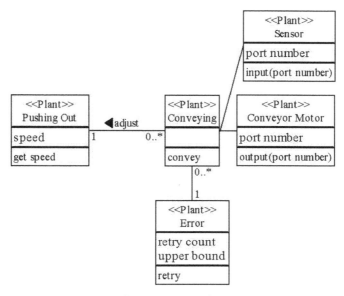

Figure 14.16 Class Diagram for Error Recovery

An evolution of conveyance is described by the sequence diagram as shown in Fig.14.17. The Controller conveys a part by the Sorting, then the Rotating rotates the conveyor and the Conveying conveys the part. During the conveyance, if an error occurs, the Error retries to convey the part and the Conveying conveys the part again. At last, notifying conveyed result is sent from the Conveying to the Sorting.

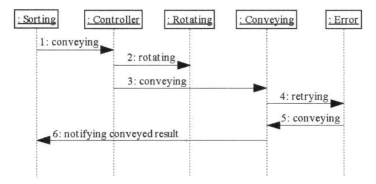

Figure 14.17 Sequence Diagram for Evolution of Conveyance

Chapter 15. From Object Orientation to Agent Orientation

Toward Next Generation Industrial System

From a historical point of view, industry was developed by a steam engine, an internal combustion engine, electric energy, information and communication technologies, etc. Manufacture is in the core of industry. Japanese manufacturing industries have a traditional base and are now being activated. In the United States of America, manufacture has highly developed by information and communication technologies. These days, research and developments of industrial systems toward the next generation are furthered by more utilization of information and communication technologies.

The development of information and communication technologies has brought about the structural change of economic social activities by the increase of each individual mutual relevance. In order to go along with such times, we should develop industrial systems which deal with product's whole life cycle such as factory automation, a business entity, consumer behavior, resources reuse, low environmental load, abandonment processing, etc.

A scalable, agile and flexible system architecture with communication and cooperation mechanisms can adapt to social changes. Such a system can be realized by dynamically connected structure of autonomous elements. This chapter discusses this kind of system which is called a **multi-agent system**[1]. It serves as a next generation industrial system[2].

Multi-agent System

A multi-agent system is a system modeled as a collection of **autonomous agents**[3]. Each agent performs a limited and local task, and interacts with a few other agents. The overall architecture of connections among agents defines the application of a system. This research has been done in the field of **distributed artificial intelligence**[1].

Nowadays, **agent orientation** is widely recognized as a new paradigm for the application of a system[4]. For example, there is a multi-agent system for information search on the World Wide Web. In this system, an agent is sent to a Web server at first. Then, the agent searches information according to some given procedures. Also, a copy of an agent is occasionally sent to a Web server if necessary. The searched results by agents are arranged on the multi-agent system. At last, their data are replied to a user. The advantage of this way is that the network traffic of a multi-agent system is lower than that of direct search by a user.

If S is a set of states of an environment, P is a set of inputs under observation of an environment, A is a set of possible actions by agents and $env : S \times A \to S$ is a function which indicates changing an environment by an action of each agent, then

[Definition 15.1] Multi-agent systems are defined by[1]
$$MA = (I, see, action, next, i_0)$$
where I is a set of internal states of an agent. $see : S \to P$ is a function which indicates observing an environment. $action : I \to A$ is a function which determines an action. $next : I \times P \to I$ is a transition function for internal states. i_0 is an internal initial state.

A modeling language which describes such sets and functions should be introduced in order to realize the MA as a real system. Then, we consider what on earth an agent is at first and next how to model an agent by a modeling language. And after that, multi-agent modeling will be discussed.

What is an Agent?

An agent is naturally implemented by an object. There are two kinds of object: an active and a passive object. What is the difference between an agent and an active object? When an active object receives a message for its operation, the operation is executed. On the other hand, if such a message is sent to an agent, the agent can reject the message. This is why an agent is also said to be an **autonomous agent**[3].

It is difficult to understand the emergent behavior of autonomous agents. Petri nets are useful for the complexity of multi-agent systems. We consider **object-oriented Petri nets**[5] as an extended class of Petri nets, and an extended Petri net for a multi-agent system will be presented at last.

Object-oriented Modeling

An object-oriented Petri net is described as in Fig.15.1. The place P_1 in the net PN1 has the net PN2 and PN3 as tokens. PN1 is a system environment, and PN2 and PN3 are objects. The symbol x on the arc from P_1 to T_1 is reference to either PN2 or PN3. When T_1 fires, either PN2 or PN3 which is referred by the x is shifted, especially, the notation "x: s1" in T_1 implies message passing where the transition T_3 or T_6 with the operation s1 fires in PN2 or PN3 which is referred by the x. This is called **reference semantics**. On the other hand, a token in P_3 is also shifted according to general firing rules. This is called **value semantics**.

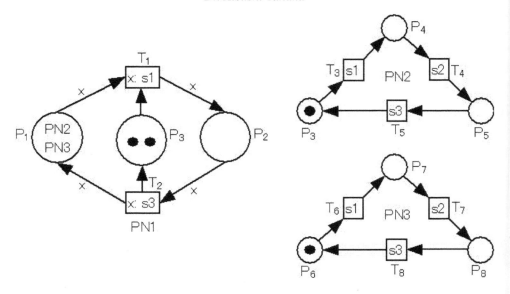

Figure 15.1 Object-oriented Petri Net

For example, we consider behavior in case where PN2 is referred by the x and T_1 fires. PN2 is taken away from P_1 and is added into P_2. A token is taken away from P_3. In PN2, a token is taken away from P_3 and is added into P_4. Thus, the Petri nets of Fig.15.2 are obtained.

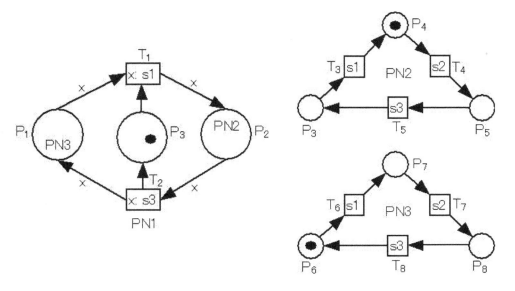

Figure 15.2 Message Passing from PN1 to PN2

When T_1 fires again, the Petri nets of Fig.15.3 are obtained.

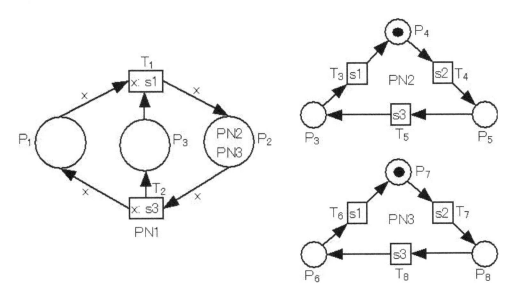

Figure 15.3 Message Passing from PN1 to PN3

Next, when the transitions T_4 and T_7 with the common operation s2 in PN2 and PN3 synchronously fires, a token is taken away from P_4 and P_7, and is added into P_5 and P_8. Thus, the Petri nets of Fig.15.4 are obtained. Moreover, if T_2 fires twice, the Petri nets return to the initial states.

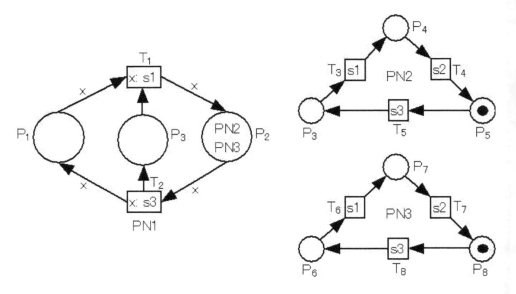

Figure 15.4 Synchronous Firing

Agile Modeling

The results of software development are of course made up of program codes. Thus, the methodology requirements for a user or a system agilely reflects program codes which deserves our attention[6]. Modeling based on program codes is called **agile modeling**.

For example, a Petri net which expresses a program code of a simple Auto Guided Vehicle (AGV) system is shown in Fig.15.5. When T_1 fires, a token is taken away from P_1, a token which is referred by the x is added into P_2, and a Petri net which indicates an object AGV is created by the notation "x: new AGV".

This operation "new" is said to be a **constructor**. Thus, the Petri net of Fig.15.6 is obtained.

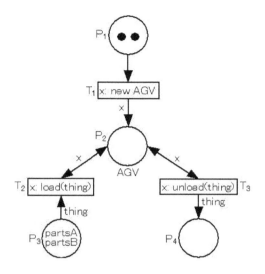

Figure 15.5 Agile Modeling for Simple AGV System

The token in P_2 corresponds the Petri net consisting of T_4, T_5, P_5 and their connected arcs. Namely, one AGV indicated by the right side net exists in the system environment indicated by the left side net. T_2 has the notation "x: load(thing)" which implies message passing to the operation "load(thing)" of T_4. When T_2 fires, the "partsA" is added into P_5 if it is taken away from P_3. Then, the Petri nets of Fig.15.7 are obtained.

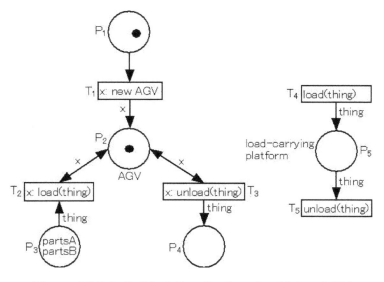

Figure 15.6 Agile Modeling for Creating Object AGV

The Petri nets of Fig.15.7 imply that the "partsA" is loaded onto the load-carrying platform of the AGV. T_3 has the notation "x: unload(thing)" which implies message passing to the operation "unload(thing)" of T_5. When T_3 fires, the "partsA" is taken away from P_5. Then, the Petri nets of Fig.15.8 are obtained. The Petri nets of Fig.15.8 imply that the "partsA" is unloaded from the load-carrying platform of the AGV.

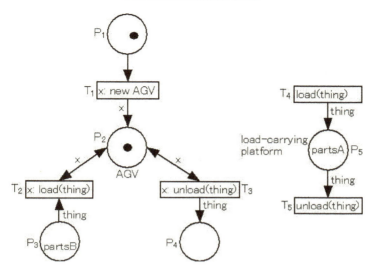

Figure 15.7 Agile Modeling for Message Passing of load(partsA)

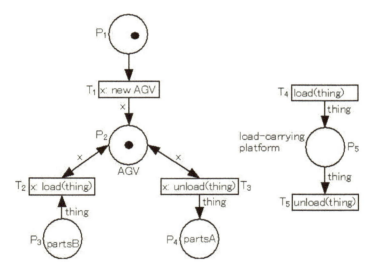

Figure 15.8 Agile Modeling for Message Passing of unload(partsA)

Multi-agent Modeling

We consider the system architecture of an industrial system by example of the AGV system environment shown in Fig.15.9[4]. There are AGV1, AGV2, and AGV stations – A and F at Zone 1, E at Zone 2, B at Zone 3, and C and D at Zone 4.

201

AGV1 goes from A to D and returns to A, and AGV2 goes from C to F and returns to C. Each AGV takes any routes, however, there is a control specification where each zone has at most one AGV.

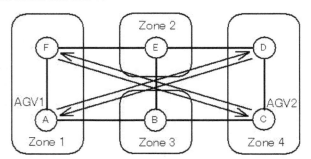

Figure 15.9 AGV System Environment

Each AGV is regarded as an agent and is naturally expressed by an object. In accordance with the UML fashion, the dynamics are described by a state machine. AGV1 and AGV2 are described by the state machines shown in Fig.15.10 and 15.11, respectively.

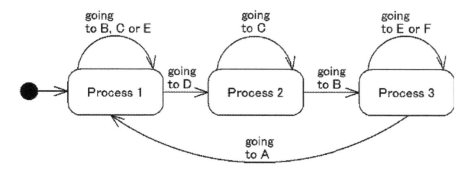

Figure 15.10 State Machine for Agent AGV1

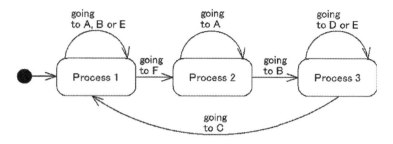

Figure 15.11 State Machine for Agent AGV2

A supervisor in Zone 1 and Zone 4 is described by the state machine shown in Fig.15.12. These zones have AGV1 and AGV2 initially, thus the initial state of the state machine is "Supervision".

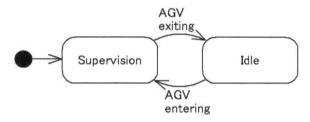

Figure 15.12 State Machine for Supervisor in Zone 1 and 4

A supervisor in Zone 2 and Zone 3 is described by the state machine shown in Fig.15.13. These zones are empty initially, thus the initial state of the state machine is "Idle".

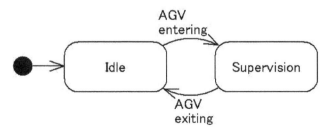

Figure 15.13 State Machine for Supervisor in Zone 2 and 3

Dr. Atsushi Tanaka

When an agent is regarded as an object and its dynamics are described by a state machine, the dynamics of a multi-agent system are expressed by an object-oriented Petri net. The dynamics of AGV1 and 2 are described by the sub Petri net shown in Fig.15.14. Each place expresses an AGV station. Then, the places – A and C have initial tokens.

As an example, the transition T_1 has the notations – "x: go to B", "y: exit" and "z: enter". These notations are interfaces for the action of objects, which are implied as follows: An AGV, which is referred by the x, goes to B; The state of a supervisor, which is referred by the y, at Zone 1, becomes "exit"; The state of a supervisor, which is referred by the z, at Zone 2, becomes "enter". The objects can be executed only if all relevant interfaces are satisfied. Thus, multi-agent system architecture is found to be specified by an object-oriented Petri net with object reference interfaces.

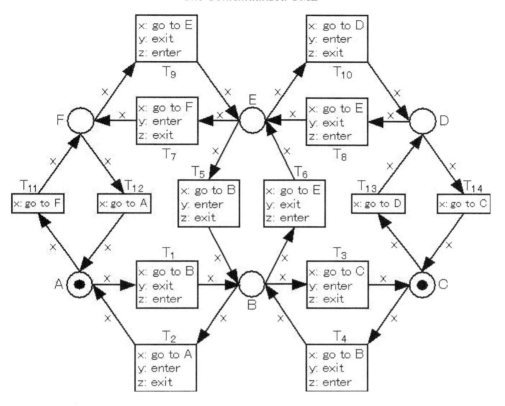

Figure 15.14 Sub Net for AGV in Multi-agent System Environment

Similarly, sub Petri nets for supervisors at Zone 1, 2, 3 and 4 are shown in Fig.15.15, 15.16, 15.17 and 15.18, respectively.

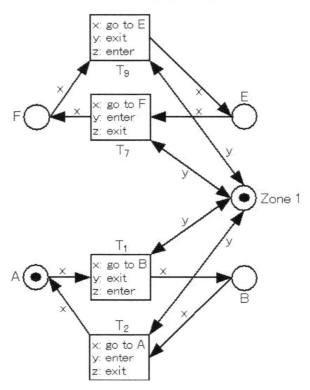

Figure 15.15 Sub Net for Supervisor at Zone 1 in Multi-agent System Environment

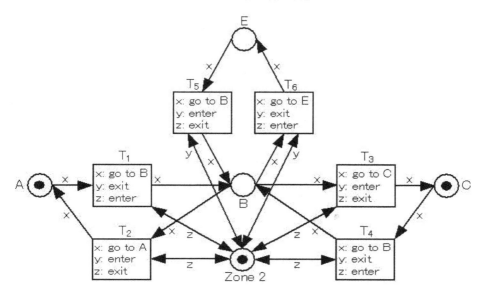

Figure 15.16 Sub Net for Supervisor at Zone 2 in Multi-agent System Environment

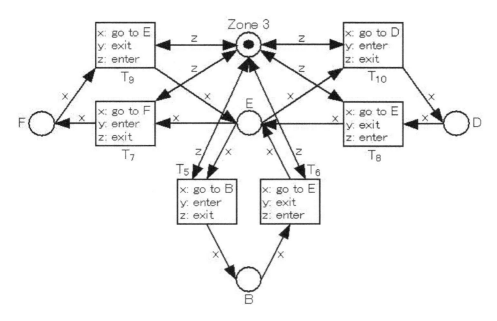

Figure 15.17 Sub Net for Supervisor at Zone 3 in Multi-agent System Environment

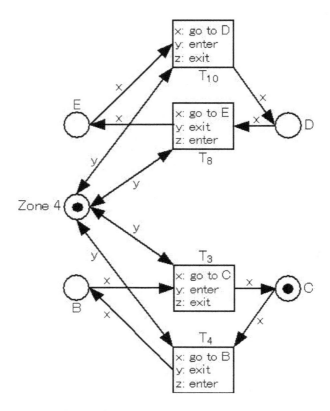

Figure 15.18 Sub Net for Supervisor at Zone 4 in Multi-agent System Environment

When we consider complexity of system dynamics, the size of a reachable set is important. For example, the dynamics of AGV1 are described by the finite automaton shown in Fig.15.19. As compared the automaton of Fig.15.19 with the state machine of Fig.15.10, the state machine is more compact. Therefore, the multi-agent modeling presented in this chapter is found to be useful.

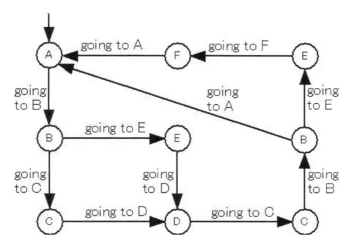

Figure 15.19 Finite Automaton for AGV1

ENDNOTES

Chapter One

1. Object Management Group, "UML 2.0 Superstructure Final Adopted Specification", *UML™ Resource Page (http://www.omg.org/uml/)*, Ch.16, Aug. 2nd 2003.

2. Takemasa, A., "First Learning UML", *NATSUME*, Ch.2, Dec. 2002.

3. Naiburg, E. J. and Maksimchuk, R. A., "UML for Database Design", *Addison Wesley*, Ch.3, 2001.

Chapter Two

1. Stevens, P. and Pooley, R., "Using UML: Software Engineering with Objects and Components", *Addison Wesley*, Ch.2, 1999.

2. Object Management Group, "UML 2.0 Superstructure Final Adopted Specification", *UML™ Resource Page (http://www.omg.org/uml/)*, Ch.7, Aug. 2nd 2003.

Chapter Three

1. Object Management Group, "UML 2.0 Superstructure Final Adopted Specification", *UML™ Resource Page (http://www.omg.org/uml/)*, p.131, Aug. 2nd 2003.

2. Object Management Group, "UML 2.0 Superstructure Final Adopted Specification", *UML™ Resource Page (http://www.omg.org/uml/)*, pp.444-447, Aug. 2nd 2003.

3. Object Management Group, "UML 2.0 Superstructure Final Adopted Specification", *UML™ Resource Page (http://www.omg.org/uml/)*, p.387, Aug. 2nd 2003.

4. Object Management Group, "UML 2.0 Superstructure Final Adopted Specification", *UML™ Resource Page (http://www.omg.org/uml/)*, pp.435-444, Aug. 2nd 2003.

Chapter Four

1. Object Management Group, "UML 2.0 Superstructure Final Adopted Specification", *UML™ Resource Page (http://www.omg.org/uml/)*, Ch.15, Aug. 2nd 2003.

2. Watanabe, M., Iida, S., Ishida, T., Yamamoto, S. and Asari, K., "Embedded Systems Development with UML Dynamical Models", *Ohmsha*, Ch.2, 2003.

Chapter Five

1. Object Management Group, "UML 2.0 Superstructure Final Adopted Specification", *UML™ Resource Page (http://www.omg.org/uml/)*, Ch.12, Aug. 2nd 2003.

2. Object Management Group, "UML 2.0 Superstructure Final Adopted Specification", *UML™ Resource Page (http://www.omg.org/uml/)*, pp.447-450, Aug. 2nd 2003.

3. Object Management Group, "UML 2.0 Superstructure Final Adopted Specification", *UML™ Resource Page (http://www.omg.org/uml/)*, pp.450-454, Aug. 2nd 2003.

4. Nissanke, N., "Realtime Systems", *Prentice Hall*, Ch.5, Sept. 1997.

Chapter Six

1. Object Management Group, "UML 2.0 Superstructure Final Adopted Specification", *UML™ Resource Page (http://www.omg.org/uml/)*, Ch.8, Aug. 2nd 2003.

2. Object Management Group, "UML 2.0 Superstructure Final Adopted Specification", *UML™ Resource Page (http://www.omg.org/uml/)*, Ch.10, Aug. 2nd 2003.

3. Object Management Group, "UML 2.0 Superstructure Final Adopted Specification", *UML™ Resource Page (http://www.omg.org/uml/)*, Ch.9, Aug. 2nd 2003.

4. Object Management Group, "UML 2.0 Superstructure Final Adopted Specification", *UML™ Resource Page (http://www.omg.org/uml/)*, p.131, Aug. 2nd 2003.

Chapter Seven

1. Stevens, P. and Pooley, R., "Using UML: Software Engineering with Objects and Components", *Addison Wesley*, Ch.17, 1999.

2. Cassandras, C. G., Lafortune, S. and Safortune, S., "Introduction to Discrete Event Systems", *Kluwer Academic Publishers*, Oct. 1999.

3. Tanaka, A., "Formal Models and State Feedback Control of Hybrid Dynamical Systems based on Petri Nets", *Ph.D. thesis*, The Graduate School of Engineering Science at Osaka University, Japan, p.5, 2001.

4. Tanaka, A., "Formal Models and State Feedback Control of Hybrid Dynamical Systems based on Petri Nets", *Ph.D. thesis*, The Graduate School of Engineering Science at Osaka University, Japan, p.6, 2001.

5. Murata, T., "Petri Nets: Properties, Analysis and Applications", *Proc. IEEE*, vol.77, no.4, pp.541-580, April 1989.

6. Bail, J. L., Alla, H. and David, R., "Hybrid Petri Nets", *Proc. ECC 91*, Grenoble, France, pp.1472-1477, July 1991.

7. David, R. and Alla, H., "Petri Nets and Grafcet", *Prentice-Hall*, Ch.3, 1992.

8. Alla, H. and David, R., "Continuous and Hybrid Petri Nets", *J. Circ. Syst. Comp.*, vol.8, no.1, pp.159-188, 1998.

9. Antsaklis, P. J., "A Brief Introduction to the Theory and Applications of Hybrid Systems", Special Issue on Hybrid Systems: Theory and Applications, *Proc. IEEE*, vol.88, no.7, pp.879-887, Jul. 2000.

10. Girault, C. and Valk, R., "Petri Nets for Systems Engineering: A Guide to Modeling, Verification, and Applications", *Springer Verlag*, pp.32-34, Nov. 2002.

11. Murata, T., "Petri Nets, Marked Graphs, and Circuit-System Theory", *IEEE Circuits and Systems*, vol.11, no.3, pp.2-12, Jun. 1977.

12. Balduzzi, F., Menga, G. and Giua, A., "Optimal Speed Allocation and Sensitivity Analysis of Hybrid Stochastic Petri Nets", *Proc. SMC '98*, pp.656-662, 1998.

Chapter Eight

1. Murata, T., "Petri Nets: Properties, Analysis and Applications", *Proc. IEEE*, vol.77, no.4, p.572, April 1989.

2. Billington, J., ed., "High-level Petri Nets - Concepts, Definitions and Graphical Notation", *Final Draft International Standard ISO/IEC 15909 Version 4.7.3*, May 10, 2002.

3. Bail, J. L., Alla, H. and David, R., "Hybrid Petri Nets", *Proc. ECC 91*, Grenoble, France, pp.1472-1477, July 1991.

4. David, R. and Alla, H., "Petri Nets and Grafcet", *Prentice-Hall*, Ch.3, 1992.

5. Alla, H. and David, R., "Continuous and Hybrid Petri Nets", *J. Circ. Syst. Comp.*, vol.8, no.1, pp.159-188, 1998.

6. Murata, T., "Petri Nets: Properties, Analysis and Applications", *Proc. IEEE*, vol.77, no.4, p.554, April 1989.

7. Baccelli, F., Cohen, G., Olsder, G. J. and Quadrat, J. P., "Synchronization and Linearity: An Algebra for Discrete Event Systems", *John Wiley & Sons*, p.59, 1992.

8. Silva, M. and Teruel, E., "A System Theory Perspective of Discrete Event Dynamic Systems: The Petri Net Paradigm", *Proc. IEEE*, Syst. Man Cyber. CESA'96, pp.1-12, 1996.

9. Cohen, G., Dubois, D., Quadrat, J. P. and Viot, M., "A Linear-System-Theoretic View of Discrete Event Processes", *22nd IEEE CDC*, pp.1039-1044, Dec. 1983.

10. Cohen, G., Dubois, D., Quadrat, J. P. and Viot, M., "A Linear-System-Theoretic View of Discrete Event Processes and Its Use for Performance Evaluation in Manufacturing", *IEEE Trans. Auto. Cont.*, vol. AC-30, no. 3, pp.210-220, Mar. 1985.

11. Baccelli, F., Cohen, G., Olsder, G. J. and Quadrat, J. P., "Synchronization and Linearity: An Algebra for Discrete Event Systems", *John Wiley & Sons*, p.39, 1992.

12. Murata, T., "Petri Nets: Properties, Analysis and Applications", *Proc. IEEE*, vol.77, no.4, p.546, April 1989.

13. Murata, T., "Petri Nets: Properties, Analysis and Applications", *Proc. IEEE*, vol.77, no.4, p.553, April 1989.

14. Tanaka, A., "Formal Models and State Feedback Control of Hybrid Dynamical Systems based on Petri Nets", *Ph.D. thesis*, The Graduate School of Engineering Science at Osaka University, Japan, pp.31-45, 2001.

15. The symbol Φ is generally used for reduction methods. In Japan, the Φ_{AT} is called A. Tanaka's method, because the method which is used for the performance evaluation of a timed model, is quite different from a reduction method which can be applied to an untimed model.

16. Cohen, G., Gaubert, S. and Quadrat, J. P., "Timed Event Graphs with Multipliers and Homogeneous Min-Plus Systems", *IEEE Trans. Auto. Cont.*, vol.43, no.9, pp.1296-1302, Sept. 1998.

17. Tanaka, A., "Formal Models and State Feedback Control of Hybrid Dynamical Systems based on Petri Nets", *Ph.D. thesis*, The Graduate School of Engineering Science at Osaka University, Japan, pp.57-58, 2001.

Chapter Nine

1. Ramadge, P. J. and Wonham, W. M., "Supervisory Control of a Class of Discrete Event Processes", *SIAM J. Cont. Opt.*, vol.25, no.1, pp.206-230, Jan. 1987.

2. Gohari, P. and Wonham, W. M., "On the Complexity of Supervisory Control Design in the RW Framework", *IEEE Trans. Syst. Man Cyb.; Part B: Cyb.*, vol.30, no.5, pp.643-652, Oct. 2000.

3. Tanaka, A., "Formal Models and State Feedback Control of Hybrid Dynamical Systems based on Petri Nets", *Ph.D. thesis*, The Graduate School of Engineering Science at Osaka University, Japan, pp.61-77, 2001.

4. Dijkstra, E. W., "A Discipline of Programming", *Prentice Hall*, Ch.3, 1976.

5. Tanaka, A., "Maximally Permissive Feedback of Timed Hybrid Petri Nets with External Input D-places under Partial Observation", *Proc. SICE Kansai Sec. Symp. 2001*, pp.133-136, Oct. 2001.

Chapter Ten

1. Nakagiri, N., "Ready UML Modeling", *RIC*, Ch.2, 2001.

2. Object Management Group, "UML 2.0 Superstructure Final Adopted Specification", *UML™ Resource Page (http://www.omg.org/uml/)*, p.529, Aug. 2nd 2003.

Chapter Eleven

1. Nakagiri, N., "Ready UML Modeling", *RIC*, Ch.3, 2001.

Chapter Twelve

1. Isoda, S., "Real-world Modeling Considered Harmful – An Analysis of Object-oriented Modeling Methods", *IEICE Trans.*, vol.J83-D1, no.9, pp.946-959, Sept. 2000.

2. Hirota, T., Kumagai, S. and Inenaga, K., "A Method of Designing Database Schema based on Use Cases", *Tec. Rep. IEICE*, KBSE2002-16, Dec. 2002.

3. Cheesman, J. and Daniels, J., "UML Components: A Simple Process for Specifying Component-based Software", *Addison Wesley*, Oct. 2000.

4. Morita, K., Nagase, Y. and Higuchi, H., "Introduction to Business Modeling with UML", *SRC*, pp.96-102, 2003.

Chapter Thirteen

1. Watanabe, H., Watanabe, M., Horimatsu, K. and Tomotake, K., "Embedded UML", *SHOEISHA*, 2002.

Chapter Fourteen

1. Watanabe, H., Watanabe, M., Horimatsu, K. and Tomotake, K., "Embedded UML", *SHOEISHA*, 2002.

Chapter Fifteen

1. Weiss, G., "Multiagent Systems – A Modern Approach to Distributed Artificial Intelligence", *The MIT Press*, 1999.

2. Miyamoto, T. and Kumagai, S., "Multi Agent Net Modelling and Realization of the Next Generation Manufacturing System", *J. ISCIE*, vol.45, no.8, pp.445-450, Aug. 2001.

3. Rumbaugh, J., "Beyond UML Software Development for the 21st Century", *Proc. Rational Educational Seminar 2002 OSAKA*, p.I-19, Nov. 6, 2002.

4. Hiraishi, K., "Modeling of Multi-Agent Systems by Petri Neta", *J. ISCIE*, vol.45, no.8, pp.439-444, Aug. 2001.

5. Girault, C., and Valk, R., "Petri Nets for Systems Engineering: A Guide to Modeling, Verification, and Application", *Springer Verlag, pp.*146-157, Nov. 2002.

6. Agile Alliance, http://www.agilealliance.com/.

BIBLIOGRAPHY

Agile Alliance, http://www.agilealliance.com/.

Alla, H. and David, R., "Continuous and Hybrid Petri Nets", *J. Circ. Syst. Comp.*, vol.8, no.1, pp.159-188, 1998.

Antsaklis, P. J., "A Brief Introduction to the Theory and Applications of Hybrid Systems", Special Issue on Hybrid Systems: Theory and Applications, *Proc. IEEE*, vol.88, no.7, pp.879-887, Jul. 2000.

Baccelli, F., Cohen, G., Olsder, G. J. and Quadrat, J. P., "Synchronization and Linearity: An Algebra for Discrete Event Systems", *John Wiley & Sons*, 1992.

Bail, J. L., Alla, H. and David, R., "Hybrid Petri Nets", *Proc. ECC 91*, Grenoble, France, pp.1472-1477, July 1991.

Balduzzi, F., Menga, G. and Giua, A., "Optimal Speed Allocation and Sensitivity Analysis of Hybrid Stochastic Petri Nets", *Proc. SMC'98*, pp.656-662, 1998.

Billington, J., ed., "High-level Petri Nets - Concepts, Definitions and Graphical Notation", *Final Draft International Standard ISO/IEC 15909 Version 4.7.3*, May 10, 2002.

Cassandras, C. G., Lafortune, S. and Safortune, S., "Introduction to Discrete Event Systems", *Kluwer Academic Publishers*, Oct. 1999.

Cheesman, J. and Daniels, J., "UML Components: A Simple Process for Specifying Component-based Software", *Addison Wesley*, Oct. 2000.

Cohen, G., Dubois, D., Quadrat, J. P. and Viot, M., "A Linear-System-Theoretic View of Discrete Event Processes", *22nd IEEE CDC*, pp.1039-1044, Dec. 1983.

Cohen, G., Dubois, D., Quadrat, J. P. and Viot, M., "A Linear-System-Theoretic View of Discrete Event Processes and Its Use for Performance Evaluation in Manufacturing", *IEEE Trans. Auto. Cont.*, vol. AC-30, no. 3, pp.210-220, Mar. 1985.

Cohen, G., Gaubert, S. and Quadrat, J. P., "Timed Event Graphs with Multipliers and Homogeneous Min-Plus Systems", *IEEE Trans. Auto. Cont.*, vol.43, no.9, pp.1296-1302, Sept. 1998.

David, R. and Alla, H., "Petri Nets and Grafcet", *Prentice-Hall*, 1992.

Dijkstra, E. W., "A Discipline of Programming", *Prentice Hall*, 1976.

Girault, C. and Valk, R., "Petri Nets for Systems Engineering: A Guide to Modeling, Verification, and Applications", *Springer Verlag*, Nov. 2002.

Gohari, P. and Wonham, W. M., "On the Complexity of Supervisory Control Design in the RW Framework", *IEEE Trans. Syst. Man Cyb.; Part B: Cyb.*, vol.30, no.5, pp.643-652, Oct. 2000.

Hiraishi, K., "Modeling of Multi-Agent Systems by Petri Nets", *J. ISCIE*, vol.45, no.8, pp.439-444, Aug. 2001.

Hirota, T., Kumagai, S. and Inenaga, K., "A Method of Designing Database Schema based on Use Cases", *Tec. Rep. IEICE*, KBSE2002-16, Dec. 2002.

Isoda, S., "Real-world Modeling Considered Harmful – An Analysis of Object-oriented Modeling Methods", *IEICE Trans.*, vol.J83-D1, no.9, pp.946-959, Sept. 2000.

Miyamoto, T. and Kumagai, S., "Multi Agent Net Modelling and Realization of the Next Generation Manufacturing System", *J. ISCIE*, vol.45, no.8, pp.445-450, Aug. 2001.

Morita, K., Nagase, Y. and Higuchi, H., "Introduction to Business Modeling with UML", *SRC*, pp.96-102, 2003.

Murata, T., "Petri Nets, Marked Graphs, and Circuit-System Theory", *IEEE Circuits and Systems*, vol.11, no.3, pp.2-12, Jun. 1977.

Murata, T., "Petri Nets: Properties, Analysis and Applications", *Proc. IEEE*, vol.77, no.4, pp.541-580, April 1989.

Naiburg, E. J. and Maksimchuk, R. A., "UML for Database Design", *Addison Wesley*, 2001.

Nakagiri, N., "Ready UML Modeling", *RIC*, 2001.

Nissanke, N., "Realtime Systems", *Prentice Hall*, Sept. 1997.

Object Management Group, "UML 2.0 Superstructure Final Adopted Specification", *UML™ Resource Page (http://www.omg.org/uml/)*, Aug. 2nd 2003.

Ramadge, P. J. and Wonham, W. M., "Supervisory Control of a Class of Discrete Event Processes", *SIAM J. Cont. Opt.*, vol.25, no.1, pp.206-230, Jan. 1987.

Rumbaugh, J., "Beyond UML Software Development for the 21st Century", *Proc. Rational Educational Seminar 2002 OSAKA*, Nov. 6, 2002.

Silva, M. and Teruel, E., "A System Theory Perspective of Discrete Event Dynamic Systems: The Petri Net Paradigm", *Proc. IEEE*, Syst. Man Cyber. CESA'96, pp.1-12, 1996.

Stevens, P. and Pooley, R., "Using UML: Software Engineering with Objects and Components", *Addison Wesley*, 1999.

Takemasa, A., "First Learning UML", *NATSUME*, Ch.2, Dec. 2002.

Tanaka, A., "Formal Models and State Feedback Control of Hybrid Dynamical Systems based on Petri Nets", *Ph.D. thesis*, The Graduate School of Engineering Science at Osaka University, Japan, 2001.

Tanaka, A., "Maximally Permissive Feedback of Timed Hybrid Petri Nets with External Input D-places under Partial Observation", *Proc. SICE Kansai Sec. Symp. 2001*, pp.133-136, Oct. 2001.

Watanabe, H., Watanabe, M., Horimatsu, K. and Tomotake, K., "Embedded UML", *SHOEISHA*, 2002.

Watanabe, M., Iida, S., Ishida, T., Yamamoto, S. and Asari, K., "Embedded Systems Development with UML Dynamical Models", *Ohmsha*, 2003.

Weiss, G., "Multiagent Systems – A Modern Approach to Distributed Artificial Intelligence", *The MIT Press*, 1999.

About The Author

Dr. Atsushi Tanaka has served as a researcher, developer and teacher in Japan. He has a Ph.D. in engineering from the graduate school of engineering science at Osaka university. His research and development interest includes concurrent, distributed, real-time, discrete event and hybrid dynamical systems with net theory and object-oriented formalism.